1   nominal mass of the singly charged ion i

2   elemental composition

3   proportion of this entry out of the total entries of this type

4   average abundance (maximum 99)

5   specificity of this explanation for the peak vs. other probable explanations (maximum 99)

6   proportion of data-base spectra with mass 71 of $\geq 1\%$ abundance

7   proportion of data-base spectra with an "important" peak at mass 71

8   for the substructure $Y_L$-$(CH_2)_3$-CO-$Y_R$, 70% of the $Y_L$ neighbors are $CH_2$ groups and 25% of $Y_R$ are ethers

9   proportion of mass 71 ions assigned as $C_4H_7O^+$

10  proportion of $C_4H_7O^+$ assignments from compounds containing -$(CH_2)_3$-CO-

11  "+" indicates that the proportion for compounds of molecular weight >236 is larger by >25% (absolute) vs. the proportion for others

12  butyryl or isobutyryl

13  no substructures indicated for hydrocarbons because of high tendency to rearrange

14  substituent at each end of substructure

15  -O-CH-CH-O-
    $H_2C$
        (cyclic)

16  most common substructures yielding $C_4H_9N^+$; data are averages for all

17  e.g., pyrollidinyl

18  less common compositions found for mass 71 peaks

19  methoxycycloalkyl

    formation of the indicated ion can involve H transfer

20  to the substructure shown, and can also involve H transfer

21  from the substructure shown

# Mass Spectral Correlations

# Mass Spectral Correlations

## Second Edition

### Fred W. McLafferty

*Department of Chemistry*
*Cornell University*
*Ithaca, New York 14853*

### Rengachari Venkataraghavan

*Department of Chemistry*
*Cornell University*
*Ithaca, New York 14853*
*Current affiliation:*
*Lederle Laboratories*
*Pearl River, New York 10965*

ADVANCES IN CHEMISTRY SERIES **40**

AMERICAN CHEMICAL SOCIETY
WASHINGTON, D. C.     1982

Library of Congress CIP Data

McLafferty, Fred W.
  Mass spectral correlations.
  (Advances in chemistry series, ISSN 0065–2393; 40)

  Bibliography: p. 10

  1. Mass spectrometry.
  I. Venkataraghavan, Rengachari, 1939–   . II. Title.
III. Series.

QD1.A355 no. 40, 1982   [QD96.M3] 540s   81–20564
ISBN  0–8412–0702–X     [543'.0873]       AACR2
                    ADCSAJ 40 1–124 1982

# Advances in Chemistry Series

**M. Joan Comstock,** *Series Editor*

# FOREWORD

ADVANCES IN CHEMISTRY SERIES was founded in 1949 by the American Chemical Society as an outlet for symposia and collections of data in special areas of topical interest that could not be accommodated in the Society's journals. It provides a medium for symposia that would otherwise be fragmented, their papers distributed among several journals or not published at all. Papers are reviewed critically according to ACS editorial standards and receive the careful attention and processing characteristic of ACS publications. Volumes in the ADVANCES IN CHEMISTRY SERIES maintain the integrity of the symposia on which they are based; however, verbatim reproductions of previously published papers are not accepted. Papers may include reports of research as well as reviews since symposia may embrace both types of presentation.

# CONTENTS

# INTRODUCTION

# INTRODUCTION

Since the First Edition of this book appeared in 1963 (1), mass spectrometry has become a widely accepted technique for molecular structure determination. Particularly impressive is the extensive use of gas chromatography/mass spectrometry resulting from its unique analytical applicability to complex mixtures. Identification of scores of components, even at the subnanogram level, is possible, but requires interpretation of the individual spectra. Literally thousands of papers have now appeared correlating mass spectra with structure for a wide variety of compounds (2), but these emphasize the spectral patterns or decomposition pathways to be expected for a specific type of molecular structure. However, in determining the structure of an unknown compound the situation is reversed; it is from the known prominent ions in the unknown spectrum that the probable structures must be ascertained. This similar problem in other fields of spectroscopy has led to charts or tables indicating the prominent functional group or other structural features which are found at particular wavelengths. Possibly the best known is the "Colthup chart" (3) of infrared spectroscopy, whose wide utility led to the original suggestion for this tabulation (4).

Extended use of the First Edition has led to a number of suggestions for improvements as well as additional correlations. A real motivation was supplied by the availability of a reference file containing mass spectra of ten times as many compounds as the original file. Correlating these spectra was only possible with computer assistance, but this had the advantage of yielding much more extensive and accurate statistical data. This compilation lists more than 3,000 structures corresponding to 1,500 elemental compositions, several times the numbers of the First Edition.

# MASS SPECTRAL CORRELATIONS

## Format of the Correlations

The tabulation lists the most probable elemental compositions and substructures corresponding to specific m/z values of singly charged ions found in electron ionization (~70 eV) mass spectra. Under each m/z value (starting at 12) are listed the most probable elemental compositions and their exact masses; for most compositions the common substructure assignments are given, and for some of these common neighboring groups are listed. Other data are the proportion of total entries represented by a specific entry and the weighted average of peak abundances for the entry. The degree of ambiguity in classifying the entry is indicated by the value for its specificity. Most entries resulted from the computer-assisted procedure described below, after which other correlations (such as peaks from skeletal rearrangements) were added using the First Edition and other tabulations (2).

## Preparation of the Correlations

These data have been taken from a collection of electron-ionization mass spectra of 32,830 different compounds (5) whose structures are coded in Wiswesser Line Notation (WLN) (6). The correlations utilized a DEC PDP-11/45 computer system with GT-40 CRT display and DIVA 58 Mbyte disk system. The spectral data with the compound name, molecular formula, molecular weight, and WLN were stored on the disk as one file. To facilitate structure manipulation, the WLNs were decoded into a connection table showing the individual units and their connections to other units using a computer program similar to the one described by Hyde et al. (6). The WLN connection table preserves the linear connectivity information, and so is particularly useful in assigning specific substructures to fragment ions. The computer-assisted correlation of spectral peaks with structure involves four major steps: the selection of significant peaks from individual spectra, assignment of all possible elemental compositions, assignment of substructures, and statistical tabulation of the results. The following is a brief description of each step in the process.

Selection of Significant Peaks. The abundances of major peaks in each spectrum are first corrected for isotopic contributions estimated from the elemen-

tal formula of the compound. Using the peak selection
procedures developed for the Probability Based Matching System (7), the peaks are then assigned uniqueness
(U) and abundance (A) values; these are based on the
occurrence probability of peaks in the data base (5),
shown in Figure 1. From each spectrum the 15 to 26
peaks (more for compounds of higher molecular weight)
of highest (U + A) values are selected (7) as the "condensed spectrum" and written on a separate disk file.

Assignment of Elemental Compositions. This program uses the molecular formula as the upper bound to
determine the numbers and kinds of elements possible
in each peak of the condensed spectrum. Improbable
compositions exceeding bonding limitations (e.g.,
$C_2H_8^+$) or with a much higher degree of unsaturation
than the molecule (e.g., $C_7H^+$ from $C_8H_{18}$) are elimin-
ated to generate the list of possible elemental compositions. These data are mapped according to their
heteroatom content so that ions with specific heteroatom compositions are grouped together. For example,
an organic compound with one oxygen and one nitrogen
atom will have four groups: one with only carbon and
hydrogen atoms, one with these plus an oxygen, one
with these plus a nitrogen, and one with these and an
oxygen and a nitrogen. This step facilitates the substructure assignment by enabling the programs to start
from specific heteroatom centers in the molecule and
determine all possible substructures for all ions with
the same heteroatom compositions.

Assignment of Substructures. Because of the high
tendency for rearrangement accompanying the formation
of hydrocarbon ions (2), specific substructures are
assigned only to peaks with elemental compositions
containing one or more heteroatoms. Given a particular heteroatom composition, the program labels their
locations in the connection table description of the
structures. Starting from each identified heteroatom
location, a neighbor unit connected to it is added
and the elemental composition comprising the units
computed. This composition is compared against the
assigned composition; if it is identical (the number
of hydrogen atoms is allowed to differ by $\pm2$) the substructure is stored as a possible assignment for the
composition. In addition, all the neighbor units connected to the terminals of the substructure are saved.
All paths from the same heteroatom center in the
molecular graph are explored to assign all possible
substructures. This process is repeated starting

5

from all other heteroatom centers in the structure,
if any, for a particular heteroatom composition. When
a composition has more than one heteroatom, all possible combinations of heteroatom centers in the structure are considered by taking them one at a time in
assigning substructural possibilities.

Scoring of Substructure Probabilities. This procedure often results in multiple substructure assignments for a possible elemental composition of a peak.
For heteroatom-containing ions, the probability that
an assignment is correct is estimated by a scoring
system whose rules are summarized in Table I. These
rules, based on the most common types of ion fragmen-
tations (2), compare the difference in number of hydrogen atoms between the assigned elemental composition and substructure, the type of ion (odd- or even-
electron), the type of bond cleavage, the number of
bonds cleaved, and the bond environment. The two
substructures with the highest score are retained as
the most probable assignments to the fragment.
For hydrocarbon ions, the compositions, but not
structures, were correlated; the score for each is
based on a comparison of its composition and rings-
plus-double-bonds ($\underline{r}$ + $\underline{db}$) value (2) with that of the
largest hydrocarbon fragment in the structure. A
score of full, half, or eighth credit is assigned if
the $\underline{r}$ + $\underline{db}$ value of the proposed composition is less
than that of the largest hydrocarbon fragment by
≤1.5, ≤2.5, or >2.5, respectively. Further, the neutral lost from largest hydrocarbon fragment in form-
ing the ion is considered; if the ratio of the $\underline{r}$ + $\underline{db}$
value of the neutral to its number of carbons is
0.57-0.66, this score is halved; it is one-quarter if
this ratio is >0.66. All scores are further halved
if the $\underline{r}$ + $\underline{db}$ value of the fragment ion is less than
4. Only the hydrocarbon composition of highest score
is retained.

Tabulation of Results. The file for each $\underline{m}$/$\underline{z}$
value containing the results of previous steps (rel-
ative abundance, elemental composition, substructures,
and neighbor units) is read to compile the following
correlations: the occurrence of a significant peak
at the particular $\underline{m}$/$\underline{z}$ value as a percentage of the
total number of spectra examined; the occurrence of
a particular elemental composition as a percentage of
the total number of significant peaks observed at
that $\underline{m}$/$\underline{z}$ value; the occurrence of a particular substructure as a percentage of the total number of iso-

Table I. Pathway Probabilities for Heteroatom Ion Formation

| H atom loss[a] | Ion type[b] | Formed by | Score |
|---|---|---|---|
| 0 | $OE^{+\cdot}$ | Cleavage of 2 ring bonds | Full |
| 0 | $OE^{+\cdot}$ | Other | Half |
| 0 | $EE^{+}$ | α-cleavage[c] | Full |
| 0 | $EE^{+}$ | Other | Half |
| ±1 | $OE^{+\cdot}$ | Any | Full |
| ±1 | $EE^{+}$ | Only one bond cleaved | Half |
| +1 | $EE^{+}$ | α-cleavage plus a second α-cleavage or cleavage of a heteroatom bond | Full |
| +1 | $EE^{+}$ | Other | Half |
| −2 | $OE^{+\cdot}$, $EE^{+}$ | Only one bond cleaved | Quarter |
| ±2 | $OE^{+\cdot}$, $EE^{+}$ | Other | Half |

[a] Number of hydrogen atoms in the assigned composition minus the number in the possible substructure.
[b] $OE^{+}$, $EE^{+}$: odd- and even-electron ions.
[c] For a carbon attached to O, N, S, or P, cleavage of another bond to that carbon.

mers observed with that elemental composition; and the
occurrence of a particular neighbor at a particular
terminal of a substructure as a percentage of all
neighbors found at that terminal. The specificity of
an elemental composition or substructure assignment
is the reciprocal of the number of possible assignments stored; the specificity of an entry is then the
average of these individual values.

## Explanatory Notes

For each $\underline{m}/\underline{z}$ value the most common elemental composition assignments of singly-charged non-metastable
ions are listed. For each of these the most probable
substructure assignments are illustrated, and for some
common substructures the common neighboring groups are
listed. In contrast to the First Edition (loc. cit.),
the mechanism for fragment ion formation is now shown.
The entries are ranked according to the product of
their proportion and abundance values. None of the
listings is exhaustive; the entries only indicate the
most probable assignments. Structures in parentheses
are illustrative of the preceding entry; "etc" indicates that isomeric ions are commonly formed by similar pathways. The user must remember that there is a
finite possibility that the correct assignment for a
peak in an individual unknown spectrum is not represented in this compilation.
        The number of hydrogen atoms in the listed sub-
structure may actually differ from that of the indicated elemental composition; the computer correlation
considered that the rearrangement of as many as two
hydrogen atoms to or from the substructure during its
formation was possible.

Proportion: The percent value in parenthesis
following the nominal $\underline{m}/\underline{z}$ heading indicates the proportion of reference spectra in the data base which
have a peak at this nominal mass of abundance equal
to or greater than 1% (Figure 1). The value on the
same line in the Proportion column is the percentage
of reference spectra having a peak at this mass whose
abundance and "uniqueness" were sufficiently significant to be used in the correlations (those peaks
selected as significant by the Probability Based
Matching algorithm - see above). For the elemental
composition subheadings (those followed by an exact
mass value) the entry in the Proportion column indicates the percentage of the entries of this nominal
mass which were determined to have this elemental

composition. For the substructure entries (non-under-lined data in the Proportion column) the values are the percentage of the elemental composition entries which were assigned to the particular substructure shown. The percentage values following the colon after the substructure indicate the entries for a particular "neighbor" group (see below) adjacent to the substructure relative to the total number of neighbor entries at the designated location of that substructure. All these proportion values have been adjusted for multiple assignment possibilities (see Specificity below) so that the total of all entries (including those not listed here) should equal 100%.

Neighbors: Following the colon after each sub-structure are listed the most abundant neighbors in descending order of proportion; the percentage values are given only if the data were statistically signifi-cant (less accurate values are rounded to the nearest 5%). The horizontal dashes in the substructure indi-cate bonds; those which are incompletely substituted are the positions holding neighboring groups. The symbol "(-)" indicates a free bond to an undesignated neighboring group from the immediately preceding atom not in parentheses; this symbol at the left of the substructure (or following "cyc") is a free bond to the preceeding group. The symbol "$(-)_2$" indicates two such single bonds, not a double bond, which is indicated as "=". The neighbors for each of these positions, left to right, are listed together, separ-ated by semicolons; the proportion of a particular pair of neighbors occurring simultaneously is indi-cated by neighboring groups separated by a colon, listed at the end of the neighbor data. Thus "$-CH_2-CO-$: $CH_2$ 50%, CH 25%; $-O-$ 40%; $CH_2$:$-O-$ 25%" indicates for the $-CH_2-CO-$ substructure that 50% occur as $-CH_2CH_2-CO-$, 25% as $-CH(-)CH_2-CO-$, 40% as $-CH_2-CO-O-$, and 25% as $-CH_2CH_2-CO-O-$.

Abundance: The average (weighted for Specificity, below) of the abundances of the peaks are given as a percentage value in the second column.

Specificity: A particular peak can have more than one assignment of both elemental composition and substructure identity; for example, $\underline{m/z}$ 43 in $C_3H_7-CO-CH_3$ could be $C_3H_7^+$ and/or $C_2H_3O^+$, or $C_2H_3O^+$ in $CH_3-CO-OCH=CH_2$ could be $CH_3-CO^+$ or $CH_2=CHO^+$. The specificity is 100% if only one assignment is made, 50% for each assignment if two are made, and so forth.

The Specificity column shows the average percentage of the assignments for the indicated entries.

High molecular weight data (+ and - signs): The statistics were taken in two sets to ascertain the effect of molecular weight on the results. If the proportion or abundance values for the compounds of molecular weight above 236 were more than 25% (absolute) greater than those of the lower molecular weight set, a "+" follows the weighted average shown; if the value for the lower molecular weight set is more than 25% (absolute) greater than the higher, a "-" follows the weighted average value.

Acknowledgments. R. S. Gohlke, V. J. Caldecourt, and N. Wright made vital contributions to the origination of this book. Many people contributed to the Second Edition; particularly helpful suggestions came from K. Biemann, G. A. Junk, H. E. Dayringer, and K. S. Haraki. We especially thank In Ki Mun for ideas, programming assistance, and computer production of all the correlations data. Lorrene Lawrence helped develop the format and typed the entire book. Financial support for research that made this correlation possible was provided by the National Institutes of Health, and funds for the computer used came in part from the National Science Foundation.
The book is dedicated to Tibby McLafferty and Usha Venkataraghavan, who continue to provide invaluable support and encouragement.

## Literature Cited

1. McLafferty, F. W., "Mass Spectral Correlations," Advances in Chemistry Series No. 40, American Chemical Society, Washington, DC (1963).
2. McLafferty, F. W., "Interpretation of Mass Spectra," Third Edition, University Science Books, Mill Valley, CA (1980).
3. Colthup, N. B., J. Opt. Soc. Am. (1950), 40, 397.
4. McLafferty, F. W. and Gohlke, R. S., Anal. Chem. (1959), 31, 1160.
5. Stenhagen, E., Abrahamsson, S., and F. W. McLafferty, "Registry of Mass Spectral Data," extended version on magnetic tape, Wiley, New York (1978).
6. Hyde, E., Mathews, F. W., Thomson, L. H., and Wisswesser, W. J., J. Chem. Doc. (1967), 7, 200.
7. Pesyna, G. M., Venkataraghavan, R., Dayringer, H. E., and McLafferty, F. W., Anal. Chem. (1976), 48, 1362.

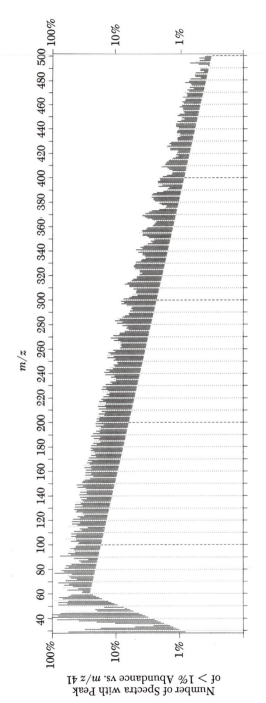

*Figure 1. Mass spectra of 29,041 different compounds containing only the elements H, C, N, O, F, Si, P, S, Cl, Br, and I in natural isotopic abundances.*

# MASS SPECTRAL DATA

| m/z, comp | Substructure, neighbor | Prop | Abnd | Spcf |
|-----------|------------------------|------|------|------|

m/z 12

C, 12.0000 small molecules

m/z 13

CH, 13.0018 small molecules

m/z 14

$CH_2$, 14.0156

N, 14.0031

m/z 15

$CH_3$, 15.0235 $CH_3$-Y*

m/z 16

O, 15.9949

$H_2N$, 16.0187

m/z 17

HO, 17.0027

$H_3N$, 17.0265

m/z 18

$H_2O$, 18.0106 data not meaningful, as water is ubiquitous in inlet systems

## MASS SPECTRAL CORRELATIONS

| m/z, comp | Substructure, neighbor | Prop | Abnd | Spcf |
|---|---|---|---|---|

### m/z 19

F, 18.9984

$H_3O$, 19.0184

### m/z 20-23 uncommon

### m/z 24

$C_2$, 24.0000 highly unsatd hc

$B_2H_2$, 24.0343

### m/z 25

$C_2H$, 25.0078 highly unsatd hc

$B_2H_3$, 25.0421

### m/z 26

$C_2H_2$, 26.0156 ar/unsatd hc

CN, 26.0031 NC-R, $RCHN_2$

$B_2H_4$, 26.0499

### m/z 27

$C_2H_3$, 27.0235 $CH_2$=CH-Y*, other hc

$CH_4B$, 27.0406 $CH_3BH$-, higher boron alkyls

$B_2H_5$, 27.0577

| m/z, comp | Substructure, neighbor | Prop | Abnd | Spcf |
|-----------|------------------------|------|------|------|

**m/z 28**

$N_2$, 28.0061  $N_2$ gas is a common contaminant

$C_2H_4$, 28.0313  hc

$CH_2N$, 28.0187  aziridines, $(CH_3)_2N-$, other amines

CO, 27.9949  lactones, etc

| m/z 29 (34%) | | 2% | 51% | |
|--------------|--|----|-----|--|
| CHO, 29.0027 | | 40 | 63 | 73 |
| $-CH_2O-$ | | 40 | 85 | 60 |
| $C_2H_5$, 29.0391 | | 17 | 60 | 75 |
| $CH_3N$, 29.0265 | | 15 | 38 | 75 |
| $-CH_2N(-)-$, $-CH_2NH-$ | | 40 | 40 | 80 |

| m/z 30 (15%) | | 5% | 25% | |
|--------------|--|----|-----|--|
| $CH_4N$, 30.0343 | | 28 | 33 | 74 |
| $-CH_2NH-$: $CH_2$ 50%; C=O 30%, $CH_2$ 30% | | 25 | 50 | 65 |
| $H_2NCH_2-$: $CH_2$ 70%, CH 17% | | 15 | 70 | 75 |
| $CH_2O$, 30.0105  $CH_3O-$, $-CH_2O-$, $HOCH_2-$ | | 33 | 17 | 76 |

also NO, 29.9979 ($-NO_2$, $-N(-)NO$, $-O-NO$); $H_2N_2$,
30.0216; $H_2Si$, 30.9921

| m/z 31 (19%) | | 4% | 29% | |
|--------------|--|----|-----|--|
| $CH_3O$, 31.0184 | | 48 | 32 | 80 |
| $HOCH_2-$: CH 45%, $CH_2$ 33%, C 13% | | 36 | 35 | 80 |
| $-CH_2O-$: $CH_2$ 47%, $CH_3$ 42%; C=O 61% | | 31 | 33 | 67 |

*17*

| m/z, comp | Substructure, neighbor | Prop | Abnd | Spcf |
|---|---|---|---|---|
| $CH_3O-$: C=O 59% | | 13 | 28 | 46 |
| CF, 30.9984 | | 11 | 21 | 82 |
| F-C(-)$_2$-: F, Cl | | 63 | 18 | 90 |
| F-ar-, F-C(-)= | | 24 | 32 | 40 |
| HNO, 31.0057 | | 4 | 25 | 67 |
| -O-N(-)- | | 55 | 28 | 46 |
| $CH_5N$, 31.0421 | | 3 | 18 | 85 |
| $CH_3NH-$: C=O 35% | | 40 | 20 | 75 |

m/z 32

Data unreliable because oxygen is a common contaminant
S, 31.9721; $O_2$, 31.9898; $CH_4O$, 32.0262

| m/z 33 (3%) | | 4% | 4% | |
|---|---|---|---|---|

$CH_5O$, 33.0340; HS, 32.9802; $CH_2F$, 33.0140; $H_2P$, 32.9894

| m/z 34 (2%) | | 3% | 2% | |
|---|---|---|---|---|
| $H_2S$, 33.9880 (can be impurity) | | 50 | 2 | 99 |

| m/z 35 (4%) | | 2% | 4% | |
|---|---|---|---|---|
| Cl, 34.9688 Cl: Y* | | 72 | 4 | 99 |
| $H_2NF$, 35.0171 | | 1 | 3 | 99 |

| m/z 36 (6%) | | 3% | 13% | |
|---|---|---|---|---|
| HCl, 35.9766 (can be impurity) | | 20 | 10 | 99 |

| m/z, comp | Substructure, neighbor | Prop | Abnd | Spcf |
|---|---|---|---|---|
| m/z 37 (12%) | | 2% | 9% | |
| C<u>3</u>H, 37.0078 | | 85 | 8 | 99 |
| m/z 38 (23%) | | 3% | 17% | |
| C<u>3</u>H<u>2</u>, 38.0156 unsatd hc | | 58 | 15 | 89 |
| C<u>2</u>N, 38.0030 arN, ar-N(-)-, ar-NH- | | 22 | 13 | 66 |
| m/z 39 (50%) | | 4% | 37% | |
| C<u>3</u>H<u>3</u>, 39.0235 HC≡CCH<u>2</u>-, ar, etc | | 57 | 38 | 96 |
| C<u>2</u>HN, 39.0108 arN, unsatd R-CN | | 20 | 24 | 75 |
| m/z 40 (34%) | | 4% | 19% | |
| C<u>3</u>H<u>4</u>, 40.0313 diunsatd/cyc hc | | 31 | 15 | 78 |
| C<u>2</u>O, 39.9949 ar(CO), R-CO-, etc | | 21 | 21 | 60 |
| C<u>2</u>H<u>2</u>N, 40.0186 arN, imines, unsatd amines, etc | | 17 | 19 | 58 |
| CN<u>2</u>, 40.0060 arN<u>2</u> | | 7 | 20 | 53 |
| m/z 41 (60%) | | 6% | 52% | |
| C<u>3</u>H<u>5</u>, 41.0391 CH<u>2</u>=CHCH<u>2</u>-, other hc | | 32 | 69 | 87 |
| C<u>2</u>HO, 41.0027 -CH<u>2</u>-CO-, -CH(-)-CO-, -C(-)<u>2</u>-CO-, ar-OH, etc | | 26 | 52 | 74 |
| C<u>2</u>H<u>3</u>N, 41.0265 arN, NC-CH<u>2</u>-, -CH<u>2</u>CH<u>2</u>N(-), | | | | |

| m/z, comp | Substructure, neighbor | Prop | Abnd | Spcf |
|---|---|---|---|---|
| $CH_2CH(-)NH-$ | | 13 | 38 | 65 |
| $CHN_2$, 41.0138 $arN_2$, $arN-NH_2$, etc | | 3 | 24 | 52 |
| m/z 42 (49%) | | 14% | 28% | |
| $C_2H_2O$, 42.0105 | | 25 | 24 | 62 |
| $-CH_2-CO-$: $CH_2$ 60%, $CH_3$ 15%; O 40%, $CH_2$ 30% | | 23 | 32 | 48 |
| $CH_3-CO-$: $-O-$ 30%, $CH_2$ 20%, $-N(-)-$ 15%, $-NH-$ 12% | | 15 | 15 | 52 |
| also $-CH(-)-CO-$, $CH_2CH(-)O-$, $-CH_2CH_2O-$, $=CH-CO-$ | | | | |
| $C_3H_6$, 42.0469 $CH_2=CH-CH_2-$, $cycC_3H_5-$, hc | | 22 | 22 | 75 |
| $C_2H_4N$, 42.0343 $CH_3N=CH-$, arN, cycN, aziridinyl, etc | | 16 | 31 | 60 |
| $CH_2N_2$, 42.0216 $arN_2$, $H_2N-arN$, $CH_3N=N-$ | | 5 | 35 | 56 |
| CNO, 41.9979 $-NH-CO-$, $-N(-)-CO-$, HO-arN, OCN-, etc | | 5 | 28 | 48 |
| $N_3$, 42.0090 $-NH-N=N-$, $-N(-)N=N-$ | | 1 | 31 | 69 |
| m/z 43 (59%) | | 9% | 72% | |
| $C_2H_3O$, 43.0184 | | 40 | 78 | 74 |
| $CH_3-CO-$: $-O-$ 28%, CH 19%, $CH_2$ 13%, $-NH-$ 10% | | 35 | 88 | 60 |
| $-CH_2-CO-$: $CH_2$ 60%, CO 21%; $-O-$ 53%, $CH_2$ 23% | | 21 | 77 | 60 |
| $-CH(OH)CH_2-$: $CH_2$ 67%; $CH_2$ 78% | | 10 | 72 | 65 |
| $-CH_2CH_2O-$: $CH_2$ 65%, $CH_3$ 20%; CO 53%, | | | | |

| m/z, comp | Substructure, neighbor | Prop | Abnd | Spcf |
|---|---|---|---|---|
| $CH_2$ 25-% | | 8 | 68 | 47 |
| $-CH_2C(-)_2O-$, $-CH(CH_3)O-$, $-CH_2OCH_2-$ | | 12 | 75 | 70 |
| $\underline{C_3H_7}$, 43.0547 | | 18 | 79 | 87 |
| $\underline{C_2H_5N}$, 43.0421 | | 7 | 52 | 60 |
| cycN, $-CH_2CH_2N(-)-$, $C_2H_5NH-$, | | | | |
| $-CH_2CH(-)NH-$, $-CH(CH_3)N(-)-$, | | | | |
| $-CH_2CH_2NH-$, $-CH_2N(CH_3)-$ | | 52 | 72 | 52 |
| CHNO, 43.0057 $-NH-CO-$, $-N(-)-CO-$, | | | | |
| $H_2N-CO-$ | | 5 | 56 | 61 |
| $\underline{CH_3N_2}$, 43.0295 $arN-NH_2$, $-N(-)N(-)CH_2-$, | | | | |
| $-N=NCH_3$, $-N=NCH_2-$ | | 3 | 53 | 66 |
| also $HN_3$, 43.0168; $C_2F$, 42.9984; CP, 42.9738 | | | | |
| m/z 44 (45%) | | 10% | 33% | |
| $\underline{C_2H_4O}$, 44.0262 | | 25 | 24 | 66 |
| $CH_3-CO-$: $-NH-$ 32%, $-O-$ 20%, N 17% | | 23 | 21 | 58 |
| $C_2H_5O-$, $HOCH_2CH_2-$, $-CH_2CH_2O-$, HO-cyc, | | | | |
| $H-CO-CH_2-$ | | 37 | 23 | 52 |
| $\underline{C_2H_6N}$, 44.0499 | | 13 | 43 | 63 |
| $CH_3CH(NH_2)-$, $CH_3NHCH_2-$: $CH_2$ 75% | | 12 | 88 | 75 |
| $-CH_2NHCH_2-$: $CH_2$ 50%, CH 30%; | | | | |
| $CH_2:CH_2$ 30% | | 7 | 55 | 88 |
| $(CH_3)_2N-$: $-CH=$, ar, C=O, $-S-$ | | 6 | 35 | 50 |
| cyc-$CH_2N(CH_3)-$, $-CH_2CH_2NH-$, $-CH_2CH(NH_2)-$, | | | | |
| $-CH_2CH_2N(-)-$, $-CH_2N(CH_3)-$, $H_2NCH_2CH_2-$ | | 34 | 30 | 62 |
| $CH_2NO$, 44.0135 | | 9 | 30 | 69 |
| $-NH-CO-$: ar 30%, $CH_2$ 20%; N 22%, | | | | |

## MASS SPECTRAL CORRELATIONS

| m/z, comp      Substructure, neighbor | Prop | Abnd | Spcf |
|---|---|---|---|
|     -NH- 17%, -O- 15% | 41 | 25 | 65 |
| $H_2N-CO-$:  $CH_2$ 35%, -NH- 20%, ar 15% | 22 | 43 | 79 |
| HO-arN, H-CO-NH-, -N(-)-CO-, ON-ar, | | | |
|   ON-CH= | 21 | 30 | 58 |
| $CO_2$, 43.9898 (also from thermal decomp) | 9 | 30 | 60 |
|   HO-CO-:  CH 30%, ar 25%, $CH_2$ 25% | 42 | 40 | 60 |
|   -O-CO-:  $CH_3$ 55%; CH 28%, ar 20%, | | | |
|     $CH_2$ 20% | 45 | 25 | 52 |
| $CH_4N_2$, 44.0373 | 3 | 40 | 58 |
|   $-N(CH_3)N(-)-$ | 25 | 50+ | 80 |
|   $-N(-)N(-)CH_2-$ | 17+ | 35 | 40 |
|   $CH_3N=N-$, $-N-CH=N-$, $H_2N-NHCH_2-$, $H_2N-arN$, | | | |
|     $H_2NC(-)=N-$, $cyc-CH_2N(NH_2)-$ | 30 | 45- | 60 |
| also $C_2HF$, 44.0062; $N_2O$, 44.0009; CS, 43.9724 | | | |
| m/z 45 (45%) | 9% | 30% | |
| $C_2H_5O$, 45.0340 | 22 | 36 | 66 |
|   $CH_3OCH_2-$:  CH 47%, $CH_2$ 28% | 15 | 59 | 75 |
|   $CH_3CHOH-$:  $CH_2$ 62%, CH 28%, C=O 6% | 8 | 70 | 70 |
|   $-CH_2OCH_2-$:  $CH_2$ 76%, $CH_3$ 15%; | | | |
|     $CH_2:CH_2$ 55% | 12 | 39 | 56 |
|   $-CH(CH_3)O-$:  $CH_3$ 50%, -O- 33%, $CH_2$ 42%, | | | |
|     C=O 38% | 5 | 43 | 76 |
|   $-CH_2CHOH-$, $HOCH_2CH_2-$, $HOC(-)_2CH_2-$, | | | |
|     $HOCH_2CH(-)-$ | 18 | 30 | 63 |
|   $CH_3CH_2O-$, $-CH_2CH_2O-$, $CH_3CH(-)-O-$ | 16 | 25 | 43 |
| $CHO_2$, 44.9976 | 18 | 27 | 70 |
|   -CO-O-:  $CH_2$ 40%, CH 15%, ar 15%; | | | |
|     Si 39%+, $CH_2$ 28%-, $CH_3$ 25% | 54+ | 20 | 58 |

| m/z, comp    Substructure, neighbor | Prop | Abnd | Spcf |
|---|---|---|---|
| HOOC-: CH$_2$ 50%, CH 20%, ar 17%, C 8% | 29– | 33 | 68 |
| -OC(-)$_2$O-, -OCH$_2$O-, -OCH(-)O- | 10 | 30 | 55 |
| CHS, 44.9799 | 7 | 24 | 86 |
| arS | 56 | 19 | 87 |
| -SCH$_2$-: CH$_2$ 45%, CH 25%; CH$_2$ 55%, CH 20% | 23 | 36 | 52 |
| ar-S- | 8 | 19 | 45 |
| C$_2$H$_7$N, 45.0578 (C$_2$H$_6$N$^+$ usually larger) | 4 | 18 | 81 |
| (CH$_3$)$_2$N-: CH$_2$ 30%, C=O 25%, -C= 15% | 53 | 20 | 85 |
| H$_2$NCH$_2$CH$_2$-: CH$_2$ 70%, NH 5%, CH 5% | 13 | 18 | 76 |
| H$_2$NCH(CH$_3$)-, -CH(NH$_2$)CH$_2$-, CH$_3$CH$_2$NH- | 25 | 13 | 85 |
| CH$_3$NO, 45.0215 H$_2$N-CO-, -NH-CO-, ONCH- | 3 | 22 | 63 |
| C$_2$H$_2$F, 45.0140 CHF=CH-, CH$_3$CF(-)-, -CF=CHF- | 1 | 21 | 72 |
| CH$_5$N$_2$, 45.0452 | 1 | 21 | 65 |
| m/z 46 (12%) | 7% | 9% | |
| NO$_2$, 45.9928 nitrates, ar-NO$_2$, R-ONO | 6 | 34 | 94 |
| CH$_2$O$_2$, 46.0054 | 18 | 5 | 82 |
| C$_2$H$_6$O, 46.0418 | 16 | 7 | 79 |
| CH$_2$S, 45.9880 -CH$_2$S-, etc | 7 | 10 | 75 |
| NS, 45.9754 S=N- | 1 | 27 | 78 |

| m/z, comp | Substructure, neighbor | Prop | Abnd | Spcf |
|---|---|---|---|---|

| m/z 47 (13%) | | 8% | 10% | |
|---|---|---|---|---|

| $\underline{CH_3S}$, 46.9959 $HSCH_2-$, $-CH_2S-$, $CH_3S-$ | 12 | 9 | 82 |
|---|---|---|---|

$\underline{CH_3O_2}$, 47.0133 $-OCH(-)O-$, $HO-CO-$     8   11   86
also CCl, 46.9688; $C_2H_7O$, 47.0496 ($CH_3O-Y*-CH_3$, poly-
  alcohols/ethers); $C_2H_4F$, 47.0297 ($CH_3CHF-$); PO,
  46.9687; $CH_4P$, 47.0051; COF, 46.9933 (F-CO-); FSi,
  47.9749

| m/z 48 (4%) | | 6% | 3% | |
|---|---|---|---|---|

| $\underline{CH_4S}$, 48.0037 | | 6 | 8 | 80 |
|---|---|---|---|---|
| $CH_3S-$: $CH_2$ 50%, CH 20%, C=O 8% | 40 | 20 | 73 |
| $-CH_2S-$: $CH_2$ 35%, CH 12%, $CH_2$ 40%, | | | |
|    CH 12% | 35 | 5 | 70 |

| $\underline{CHCl}$, 47.9766 | 9 | 3 | 89 |
|---|---|---|---|

| $\underline{H_2NO_2}$, 48.0084 $-ONO$, $-NO_2$ | 4 | 3 | 95 |
|---|---|---|---|

$\underline{H_2NS}$, 47.9910       1   16   70
also $CH_4O_2$, 48.0211; OS, 47.9673; HOP, 47.9765

| m/z 49 (9%) | | 5% | 6% | |
|---|---|---|---|---|

| $\underline{C_4H}$, 49.0078 | 61 | 5 | 98 |
|---|---|---|---|

| $\underline{CH_2Cl}$, 48.9844 | | 12 | 12 | 85 |
|---|---|---|---|---|
| $ClCH_2-$: $CH_2$ 35%, CH 30%, C=O 10% | 45 | 16 | 67 |
| $ClC(-)_2-$ | 20+ | 13 | 84 |
| $ClCH(-)-$: Cl 25%, $CH_3$ 20%, $CH_2$ 20% | 17 | 5 | 80 |

| $\underline{HOS}$, 48.9751 O=S(-)-: $CH_2$ 35%, O 35% | 1 | 3 | 99 |
|---|---|---|---|

| m/z, comp | Substructure, neighbor | Prop | Abnd | Spcf |
|---|---|---|---|---|

$\underline{H_2OP}$, 48.9843 O=P(-)$_2$-: CH$_3$ 40%, O 15%,

  S 15%        1    12    99

also H$_2$FSi, 48.9809; NCl, 48.9718; CH$_2$OF, 49.0089

| m/z 50 (35%) | | 5% | 17% | |
|---|---|---|---|---|

$\underline{C_4H_2}$, 50.0156 ar        63    16    95

$\underline{C_3N}$, 50.0030 arN        7    17    63

$\underline{CF_2}$, 49.9968 -CF$_2$- etc        5    13    89

$\underline{CH_3Cl}$, 49.9923        1    47    78

  ClCH$_2$-: C=O 45%, -C=C- 45%        70-   57    75

| m/z 51 (48%) | | 11% | 20% | |
|---|---|---|---|---|

$\underline{C_4H_3}$, 51.0235 ar        45    18    89

$\underline{C_3HN}$, 51.0108 arN, ar-N(-)-        21    21    68

$\underline{CHF_2}$, 51.0046        2    36    88

  CHF$_2$-: CX$_2$ 35%, C=O 30%        25    80    82

  -CF$_2$-        50    23    78

also FS, 50.9708; HFP, 50.9800

| m/z 52 (35%) | | 5% | 19% | |
|---|---|---|---|---|

$\underline{C_3H_2N}$, 52.0186 arN, ar-NH$_2$        20    15    51

$\underline{C_3O}$, 51.9949 ar-CO-, ar(C=O), ar-O-, etc   19    18    63

$\underline{C_4H_4}$, 52.0313 ar, unsatd hc        19    13    69

$\underline{C_2N_2}$, 52.0060 arN$_2$        11    20    51

# MASS SPECTRAL CORRELATIONS

| m/z, comp | Substructure, neighbor | Prop | Abnd | Spcf |
|---|---|---|---|---|
| m/z 53 (49%) | | 8% | 20% | |
| C$_4$H$_5$, 53.0391 polyunsatd hc | | 31 | 18 | 88 |
| C$_3$HO, 53.0027 ar(C=O), arO, ar-O-,<br>  cyc-CO-, cyc-O- | | 22 | 21 | 67 |
| C$_3$H$_3$N, 53.0265 arN, ar-NH$_2$, cycN, etc | | 11 | 18 | 58 |
| C$_2$HN$_2$, 53.0138 arN$_2$, cycN$_2$, etc | | 4 | 18 | 48 |
| m/z 54 (34%) | | 5% | 24% | |
| C$_4$H$_6$, 54.0469 | | 34 | 25 | 81 |
| C$_3$H$_2$O, 54.0105 ar(CO), R-CO-, etc | | 20 | 23 | 60 |
| C$_3$H$_4$N, 54.0343 NC-CH$_2$CH$_2$-, ar(NH), imines<br>  (-CH$_2$CH$_2$C(=NH)-) | | 11 | 24 | 51 |
| C$_2$H$_2$N$_2$, 54.0216 arN$_2$, etc | | 5 | 18 | 46 |

also C$_2$NO, 53.9979 (arNO); CN$_3$, 54.0090 (arN$_3$)

| m/z 55 (55%) | | 10% | 46% | |
|---|---|---|---|---|
| C$_4$H$_7$, 55.0547 H$_2$C=C(CH$_3$)CH$_2$-,<br>  CH$_3$CH=CHCH$_2$-, other hc | | 34 | 55 | 88 |
| C$_3$H$_3$O, 55.0184 | | 29 | 48 | 74 |
|   cyc-CH$_2$CH$_2$-CO-: CH$_2$ 50%, CH 15%;<br>    CH$_2$ 40%, -O- 30% | | 38 | 56 | 72 |
|   CH$_2$=CH-CO-, -CH=CH-CO- | | 6 | 53 | 68 |

  also -CH=CHCH$_2$O-, HOCH$_2$C≡C-, ar-OCH$_3$, etc

C$_3$H$_5$N, 55.0421 substd/cyc amines,

| m/z, comp | Substructure, neighbor | Prop | Abnd | Spcf |
|---|---|---|---|---|
| $NC-CH_2CH_2-$, $CN-CH_2CH_2-$, arN | | 8 | 31 | 61 |

$\underline{C_2HNO}$, 55.0057  $-CH_2-CO-N(-)-$,
   $-CH_2N(-)-CO-$, $OCN-CH_2-$, $-NHCH(-)-CO-$,
   etc

| | | 3 | 28 | 45 |
|---|---|---|---|---|

$\underline{C_2H_3N_2}$, 55.0295  $arN_2$, $arN-NHCH_3$,
   cyc/substd diamines

| | | 2 | 26 | 52 |
|---|---|---|---|---|

also $CHN_3$, 55.0168 ($arN_3$)

| m/z 56 (42%) | | 11% | 28% | |
|---|---|---|---|---|

$\underline{C_4H_8}$, 56.0626  $H-C_4H_8-Y*$, $H-C_4H_8-R-Y*$,
   $H-R-CH_2C(CH_3)=CH_2$ etc, hc

| | | 30 | 31 | 78 |
|---|---|---|---|---|

$\underline{C_3H_4O}$, 56.0262  substd/cyc ketones/ethers,
   etc

| | | 24 | 26 | 62 |
|---|---|---|---|---|

$\underline{C_3H_6N}$, 56.0499  cyc/subst amines
   ($cyc-CH_2CH_2CH(NH_2)-$)

| | | 11 | 29 | 57 |
|---|---|---|---|---|

$\underline{C_2H_2NO}$, 56.0135  cyc/subst amides
   ($-N(CH_3)-CO-$), $OCN-CH_2-$, ar-NO, etc

| | | 6 | 27 | 53 |
|---|---|---|---|---|

$\underline{C_2O_2}$, 55.9898  $-CH_2O-CO-$, $-CH_2-CO-O-$,
   $-CO-CO-$, etc

| | | 4 | 24 | 42 |
|---|---|---|---|---|

also $C_2H_4N_2$, 56.0373 ($arN_2$, $C_2H_5N(-)N(-)-$, etc); $CN_2O$,
56.0009; $CH_2N_3$, 56.0246 ($arN_3$)

| m/z 57 (48%) | | 11% | 43% | |
|---|---|---|---|---|

$\underline{C_3H_5O}$, 57.0340

| | | 27 | 40 | 64 |
|---|---|---|---|---|

   $-CH_2CH_2OCH_2-$: $CH_2$ 68%, OH 23%; $CH_2$ 72%,
   CH 18%

| | | 8 | 68 | 85 |
|---|---|---|---|---|

   $-CH_2CH_2-CO-$: $CH_2$ 35%, $-O-$ 20%; $CH_2$ 30%,

| m/z, comp | Substructure, neighbor | Prop | Abnd | Spcf |
|---|---|---|---|---|
| -O- 20% | | 13 | 35 | 50 |
| $C_2H_5$-CO-: $CH_2$ 33%, -O- 21% | | 7 | 76 | 57 |
| -$CH_2CH_2CH(OH)$- (cycloalkanols), | | | | |
| -$CH(CH_2OH)CH_2$-, cycO | | 10 | 60 | 80 |
| -$(CH_2)_3O$-, $CH_3CH(-)$-CO- | | 13 | 25 | 50 |
| | | | | |
| $\underline{C_4H_9}$, 57.0704 | | 24 | 55 | 85 |
| | | | | |
| $\underline{C_2HO_2}$, 56.9976 | | 8 | 36 | 53 |
| -$CH_2$O-CO-: $CH_2$ 35%, $CH_3$ 27%; CH 30% | | 38 | 33 | 50 |
| -$CH_2$-CO-O: $CH_2$ 45%, $CH_3$ 18%; $CH_3$ 47% | | 27 | 35 | 48 |
| -CO-CO- | | 6 | 50 | 60 |
| -C(-)$_2$O-CO-, cyc-CH(OH)-CO-, | | | | |
| -C(-)$_2$-CO-O-, cyc-CH(-COOH)- | | 10 | 65 | 45 |
| | | | | |
| $\underline{C_3H_7N}$, 57.0577 -$(CH_2)_3N(-)$-, | | | | |
| -$CH_2CH_2N(CH_3)$-, $CH_3$N=CHCH$_2$-, -$(CH_2)_3$NH-, | | | | |
| -$CH_2CH_2N(-)CH_2$-, -$C(CH_3)_2N(-)$- | | 8 | 27 | 60 |
| | | | | |
| $\underline{C_2H_3NO}$, 57.0214 | | 5 | 23 | 55 |

also $C_2H_5N_2$, 57.0451; $CHN_2O$, 57.0087; $CH_3N_3$, 57.0325;
  $C_2H_2P$, 56.9894

| m/z 58 (34%) | | 6% | 32% | |
|---|---|---|---|---|
| $\underline{C_3H_8N}$, 58.0656 | | 16 | 57 | 74 |
| $(CH_3)_2NCH_2$-: $CH_2$ 65%, CH 20% | | 31 | 95 | 86 |
| -$CH_2N(C_2H_5)$-: $CH_3$ 50%; C=O 60% | | 8 | 40 | 73 |
| cyc-$CH_2N(CH_3)CH_2$-: $CH_2$:$CH_2$ 50% | | 6 | 63+ | 70 |
| other cycN | | 12 | 38 | 55 |
| $C_2H_5NHCH_2$-, $C_2H_5CH(NH_2)$-, $(CH_3)_2C(NH_2)$-, | | | | |
| $CH_3NHCH(CH_3)$- | | 7 | 80 | 70 |
| | | | | |
| $\underline{C_3H_6O}$, 58.0418 | | 25 | 28 | 68 |

| m/z, comp | Substructure, neighbor | Prop | Abnd | Spcf |
|---|---|---|---|---|
| $CH_3$-CO-$CH_2$-: $CH_2$ 55%, C=O 25%, CH 10% | | 10 | 47+ | 75 |
| -$CH_2$-CO-$CH_2$- | | 6 | 55 | 94 |
| HO-cyc (cycloalkanols) | | 12 | 30 | 58 |
| $CH_3OCH_2CH_2$-, HO$(CH_2)_3$-, -$CH_2CH_2$-CO-, | | | | |
| $C_3H_7O$- | | 11 | 35 | 55 |
| $\underline{C_2H_2O_2}$, 58.0054 $CH_3O$-CO-, -$CH_2$-CO-O- | | 9 | 23 | 55 |
| $\underline{C_2H_4NO}$, 58.0292 $CH_3NH$-CO-, -N($CH_3$)-CO- | | 7 | 23 | 62 |
| $\underline{CH_2N_2O}$, 58.0165 -O-ar$N_2$, HO-ar$N_2$, | | | | |
| -NH-CO-N(-)- | | 2 | 25 | 61 |
| also $C_2H_2S$, 57.9880; $C_2H_6N_2$, 58.0529; $CNO_2$, 57.9928; | | | | |
| CNS, 57.9754 | | | | |

| m/z 59 (31%) | | 13% | 23% | |
|---|---|---|---|---|
| $\underline{C_2H_3O_2}$, 59.0133 | | 27 | 22 | 77 |
| $CH_3O$-CO-: $CH_2$ 50%, CH 20%, ar 9% | | 60 | 23 | 75 |
| -CO-$OCH_2$-: $CH_2$ 41%, CH 24%, ar 13%; | | | | |
| $CH_3$ 40%, $CH_2$ 35%, CH 15% | | 9 | 17 | 60 |
| -CH(OH)$CH_2O$-: $CH_3$ 70%; $CH_2$ 50%, ar 25% | | 2 | 38 | 65 |
| $\underline{C_3H_7O}$, 59.0496 | | 16 | 28 | 70 |
| -C($CH_3)_2O$- | | 8 | 45 | 60 |
| $C_2H_5OCH_2$-: $CH_2$ 30%, C=O 25%, CH 25% | | 5 | 45 | 55 |
| $C_2H_5CH(OH)$-: $CH_2$ 60%, CH 15% | | 4 | 75 | 65 |
| $(CH_3)_2C(OH)$-: $CH_2$ 55%, CH 20% | | 4 | 70 | 85 |
| -$CH_2CH_2OCH_2$-, -$CH_2CH_2CH_2O$-, | | | | |
| -CH($CH_3$)$OCH_2$- | | 15 | 20 | 55 |
| $\underline{C_2H_5NO}$, 59.0370 | | 4 | 24 | 69 |
| $H_2N$-CO-$CH_2$: $CH_2$ 80%, CH 15% | | 12 | 84 | 99 |
| $CH_3$-CO-NH-: CH 40%, $CH_2$ 30% | | 15 | 19 | 55 |

| m/z, comp | Substructure, neighbor | Prop | Abnd | Spcf |
|---|---|---|---|---|
| HO-N=CHCH$_2$- | | 2 | 90 | 99 |
| cyc-C(=N-OH)-CH$_2$- | | 4 | 25 | 99 |
| cyc-CH(NH$_2$)-CH(OH)- | | 3 | 50 | 99 |
| cyc-N(CH$_3$)-CO-, CH$_3$O-N=CH-, -CH$_2$NH-CO- | | 17 | 10 | 75 |
| | | | | |
| $\underline{C_2H_3S}$, 58.9959 | | 3 | 20 | 75 |
| cyc-S-CH(CH$_3$)-: C; -O-, S | | 23+ | 42+ | 90 |
| -CH(CH$_3$)S- | | 15 | 11 | 80 |
| -CH=CH-S- | | 4 | 23- | 65 |
| -CH$_2$CH$_2$S-, -CH$_2$SCH$_2$-, CH$_3$SCH(-)- | | 20 | 12 | 80 |
| | | | | |
| $\underline{C_3H_9N}$, 59.0734 (C$_3$H$_8$N usually larger) | | 3 | 7 | 77 |
| H$_2$NCH$_2$CH$_2$CH$_2$-: CH$_2$ 80% | | 16 | 9 | 76 |
| (CH$_3$)$_2$NCH$_2$-, C(CH$_3$)$_2$NH-, -CH$_2$N(C$_2$H$_5$)-, | | | | |
| (CH$_3$)$_2$CHNH- | | 30 | 10 | 70 |
| | | | | |
| $\underline{CHNO_2}$, 59.0006 O$_2$NCH$_2$-, -O-CO-N(-)-, | | | | |
| -O-CO-NH- | | 1 | 20 | 60 |
| | | | | |
| $\underline{C_3H_4F}$, 59.0297 cyc-CHFCH$_2$CH$_2$-, | | | | |
| cyc-CH$_2$CHFCH$_2$- | | 1 | 20 | 80 |
| | | | | |
| m/z 60 (19%) | | 6% | 20% | |
| | | | | |
| $\underline{C_2H_4O_2}$, 60.0211 | | 27 | 22 | 79 |
| HO-CO-CH$_2$-: CH$_2$ 60%, CH 12% | | 18 | 45 | 76 |
| -O-CO-CH$_2$-: CH$_2$:CH$_2$ 25% | | 11 | 22 | 74 |
| cyc-CH(OH)CH(OH)- | | 10 | 35 | 76 |
| | | | | |
| $\underline{C_3H_8O}$, 60.0575 cyc-CH(CH$_3$)CH$_2$O-, | | | | |
| cyc-CH$_2$CH(OH)CH$_2$-, etc | | 11 | 16 | 62 |
| | | | | |
| $\underline{C_2H_6NO}$, 60.0448 | | 6 | 28 | 72 |
| CH$_3$-CO-NH-: CH$_2$ 40%, CH 40% | | 35 | 24 | 77 |
| HOCH$_2$CH(NH$_2$)-: CH$_2$ 40%, C=O 40% | | 5 | 55 | 75 |

| m/z, comp          Substructure, neighbor | Prop | Abnd | Spcf |
|---|---|---|---|
| -CH(CH₂OH)NH-: C=O; C=O | 3 | 70 | 55 |
| (CH₃)₂N-O- | 4 | 55 | 95 |
| | | | |
| C₂H₄S, 60.0037 | 5 | 20 | 74 |
| -CH₂CH₂S-: CH₂; CH₂ | 25 | 38 | 73 |
| | | | |
| CHOP, 59.9765 arP | 1 | 20+ | 70 |
| | | | |
| m/z 61 (21%) | 5% | 16% | |
| | | | |
| C₂H₅O₂, 61.0289 | 21 | 16 | 80 |
| CH₃-CO-O-: CH₂ 65%, CH 30% | 17 | 13 | 60 |
| HO-CO-CH₂-, CH₃O-CO-: CH₂, CH | 20 | 21 | 73 |
| cyc-CH(OH)CH(OH)- | 13 | 18 | 85 |
| also CH₃OCH(-)O-, -OC(-)(CH₃)O- | | | |
| | | | |
| C₅H, 61.0078 | 24 | 10 | 90 |
| | | | |
| C₂H₅S, 61.0115 | 9 | 28 | 80 |
| CH₃SCH₂-: CH₂ 80% | 12 | 75 | 80 |
| -CH₂SCH₂- | 35+ | 35 | 80 |
| CH₃CH(SH)-, -CH(CH₃)S-, -CH₂CH₂S-, | | | |
| C₂H₅S- | 25 | 25 | 55 |
| | | | |
| C₂H₂Cl, 60.9844 -CH₂CHCl-, ClCH=CH-, | | | |
| Cl-cyc | 5 | 19 | 93 |
| | | | |
| C₂H₆P, 61.0207 (CH₃)₂P- (abnd 100%), etc | 1 | 70 | 64 |
| also CH₂OP, 60.9843 (-P-CO-); CH₅N₂O, 61.0400 | | | |
| (H₂N-CO-NH-); C₂H₂OF, 61.0089 (FCH₂-CO-); C₃H₆F, | | | |
| 61.0453 ((CH₃)₂CF-) | | | |
| | | | |
| m/z 62 (20%) | 5% | 10% | |
| | | | |
| C₅H₂, 62.0156 | 53 | 9 | 94 |

# MASS SPECTRAL CORRELATIONS

| m/z, comp | Substructure, neighbor | Prop | Abnd | Spcf |
|---|---|---|---|---|
| $\underline{C_4N}$, 62.0030 arN, ar-N(-)- | | 11 | 11 | 75 |
| $\underline{C_2H_3Cl}$, 61.9923 | | 4 | 16 | 90 |
| $\underline{C_2H_7P}$, 62.0285 $(CH_3)_2P-$ : -CO-; $C_2H_5PH-$ | | 1 | 50 | 72 |
| $\underline{C_2H_6S}$, 62.0193 $C_2H_5S-$ | | 2 | 13 | 79 |
| $\underline{CH_4NO_2}$, 62.0241 $H_2N-CO-O-$ | | 1 | 20 | 96 |
| $\underline{C_2F_2}$, 61.9968 | | 2 | 7 | 83 |
| $\underline{C_2H_6O_2}$, 62.0367 | | 1 | 10 | 84 |

also $CH_2OS$, 61.9829; $CH_4NS$, 62.0067; $H_2N_2S$, 61.9940;
  CHNCl, 61.9796 (Cl-arN)

| m/z 63 (35%) | | 8% | 18% | |
|---|---|---|---|---|
| $\underline{C_5H_3}$, 63.0235 | | 56 | 16 | 88 |
| $\underline{C_4HN}$, 63.0108 arN | | 10 | 16 | 72 |
| $\underline{C_2H_4Cl}$, 63.0001 | | 3 | 34 | 79 |
| $ClCH_2CH_2-$ : O 35%, C=O 15%, $CH_2$ 15% | | 38 | 42 | 65 |
| $ClCH(CH_3)-$ : C=O 30%, $CH_2$ 20%, CH 20% | | 11 | 54 | 43 |
| $\underline{COCl}$, 62.9637 Cl-CO- : O 40%, $CH_2$ 15% | | 1 | 41 | 97 |
| $\underline{CH_3OS}$, 62.9908 $-CH_2S(=O)-$, $CH_3OS-$, $CH_3S(=O)-$ | | 1 | 36 | 74 |
| $\underline{C_2HF_2}$, 63.0046 CHF=CF-, -CF(-)CF(-)- | | 1 | 27 | 90 |
| $\underline{CFS}$, 62.9708 cyc-CF(-)S- etc | | 1 | 26 | 99 |

also $CH_3O_3$, 63.0082 $(-C(-O-)_3$, $O=C(-O-)_2)$

| m/z, comp | Substructure, neighbor | Prop | Abnd | Spcf |
|---|---|---|---|---|
| m/z 64 (24%) | | 7% | 12% | |
| $\underline{C_5H_4}$, 64.0313 ar | | 46 | 9 | 82 |
| $\underline{C_4H_2N}$, 64.0186 arN, ar-N(-)-, etc | | 10 | 11 | 50 |
| $\underline{C_3N_2}$, 64.0060 arN$_2$ | | 5 | 12 | 43 |
| $\underline{C_4O}$, 63.9949 ar(C=O), ar-CO-, arO | | 5 | 9 | 50 |
| $\underline{C_2H_2F_2}$, 64.0124 -CH$_2$CF$_2$-, etc | | 2 | 26 | 84 |
| $\underline{S_2}$, 63.9448 disulfides | | 1 | 33 | 81 |
| $\underline{CH_4OS}$, 63.9986 CH$_3$O-S- | | 1 | 23 | 79 |
| $\underline{SO_2}$, 63.9622 (could be impurity) | | 1 | 21 | 99 |

also $C_2H_5Cl$, 64.0079; $HO_2P$, 63.9714; CHFS, 63.9786;
  $CNF_2$, 63.9998

| m/z, comp | Substructure, neighbor | Prop | Abnd | Spcf |
|---|---|---|---|---|
| m/z 65 (43%) | | 8% | 19% | |
| $\underline{C_5H_5}$, 65.0391 unsatd hc, ar | | 34 | 17 | 77 |
| $\underline{C_4H_3N}$, 65.0265 arN, ar-NH$_2$, ar-NH- | | 17 | 19 | 58 |
| $\underline{C_4HO}$, 65.0027 ar-CO-, ar(C=O), arO, ar-O-, ar-OH | | 13 | 18 | 58 |
| $\underline{C_3HN_2}$, 65.0138 arN$_2$, ar-N=N-, arN-N(-)-, ar-CH=N-NH- | | 4 | 23 | 48 |
| $\underline{H_2O_2P}$, 64.9792 O=P(OH)(-R)$_2$, O=P(O-)$_3$, etc | | 1 | 27 | 70 |
| $\underline{C_2H_3F_2}$, 65.0203 CH$_3$CF$_2$-, -CHF-CF(-)-, | | | | |

MASS SPECTRAL CORRELATIONS

| m/z, comp | Substructure, neighbor | Prop | Abnd | Spcf |
|---|---|---|---|---|
| ar-F$_2$, etc | | 1 | 44- | 75 |

also HO$_2$S, 64.9700; H$_2$ClSi, 64.9513; CHNF$_2$, 65.0076

| m/z 66 (29%) | | 7% | 13% | |
|---|---|---|---|---|
| C$_5$H$_6$, 66.0469 ar, CH$_3$-pyridines | | 26 | 11 | 84 |
| C$_4$H$_2$O, 66.0105 ar-CO-, ar-O-, ar-OH etc | | 17 | 10 | 62 |
| C$_4$H$_4$N, 66.0343 arN (pyrrolyl-, N-R-pyrrolyl) | | 12 | 11 | 50 |
| C$_3$H$_2$N$_2$, 66.0216 arN$_2$ (pyrazolyl), ar-CH=N-NH- | | 6 | 12 | 42 |
| C$_3$NO, 65.9979 ar(NH-CO-), arN-OH, arON etc | | 4 | 11 | 47 |
| H$_2$S$_2$, 65.9604 -S-S- | | 1 | 39 | 55 |
| C$_2$N$_3$, 66.0090 arN$_3$ etc | | 1 | 13 | 35 |
| CFCl, 65.9672 -CClF-, ClFC= | | 1 | 11 | 99 |

| m/z 67 (40%) | | 11% | 34% | |
|---|---|---|---|---|
| C$_5$H$_7$, 67.0547 polyunsatd/cyc hc | | 43 | 38 | 95 |
| C$_4$H$_3$O, 67.0184 substd ketones, furyl, ar(C-O), etc | | 17 | 35 | 73 |
| C$_4$H$_5$N, 67.0421 pyrrolyl, cyc imines, substd amines, etc | | 7 | 22 | 58 |
| C$_3$H$_3$N$_2$, 67.0295 arN$_2$ (imidazoyl, etc), cyc hydrazone | | 3 | 20 | 42 |
| C$_3$HNO, 67.0057 ar(-NH-CO-), arN-OH, arN-CO-, etc | | 2 | 20 | 42 |

m/z, comp    Substructure, neighbor    Prop Abnd Spcf
also $C_2N_3$, 67.0168 (arN$_3$); CHFCl, 66.9750 (FClCH-);
 CHOF$_2$, 66.9995 (-CF$_2$O-); ClS, 66.9412; OFS, 66.9657

| m/z 68 (33%) | | 6% | 27% |
|---|---|---|---|

$\underline{C_5H_8}$, 68.0626 cyclopentyl, cyc/unsatd hc   37   28   84

$\underline{C_4H_4O}$, 68.0262 subst/cyc -CO-/-O-   20   29   59

$\underline{C_4H_6N}$, 68.0499 subst/cyc amines/imines,
  nitriles   8   23   50

$\underline{C_3H_2NO}$, 68.0135 arN-OH, ar(NO)
  (isoxazolyl), ar-O-, ar(N-CO),
  NC-CH$_2$-CO-   4   24   42

$\underline{C_3H_4N_2}$, 68.0373 arN$_2$ (pyrimidinyl, etc),
  arN-NH$_2$, etc   4   17   49

$\underline{C_3O_2}$, 67.9898 ar-CO-O-, ar(C=O)-O-,
  substd/unsatd -CO-O-, etc   3   23   46
also $C_2N_2O$, 68.0009 (NarNH-CO-, etc); $C_2H_2N_3$, 68.0246
  (arN$_3$, arN$_2$-NH-); $C_4HF$, 68.0062; $CN_4$, 68.0120 (arN$_4$)

| m/z 69 (49%) | | 12% | 35% |
|---|---|---|---|

$\underline{C_5H_9}$, 69.0704 CH$_2$=CHC(CH$_3$)$_2$-, cyclopentyl,
  other unsatd/cyc hc   28   35   90

$\underline{C_4H_5O}$, 69.0340   20   33   66
  cyc-CH(CH$_3$)CH$_2$-CO-, cyc-CH$_2$CH(CH$_3$)-CO-,
  cyc-CH$_2$CH$_2$CH(-CO-)-, cyc-(CH$_2$)$_3$-CO-,
  etc   25+   35   72
  CH$_3$CH=CH-CO-, other C$_3$H$_5$-CO-,
    CH$_3$-CO-CH=CH-   5   55   60

MASS SPECTRAL CORRELATIONS

| m/z, comp | Substructure, neighbor | Prop | Abnd | Spcf |
|-----------|------------------------|------|------|------|

also $CH_3C\equiv CCH(OH)-$, cyc/unsatd/subst -O-/OH

$\underline{C_3HO_2}$, 68.9976 unsatd/cyc/ar
-CO-/-O-/-OH (-CH=CH-CO-, etc)      7    33    56

$\underline{CF_3}$, 68.9952 $CF_3-$, polyfluoro/haloalkanes   3    63    73

$\underline{C_3H_3NO}$, 69.0214 -CH=CH-CO-NH-, HO-arN,
OarN, ar-NO, etc        4    21    48

$\underline{C_4H_7N}$, 69.0577 $NC(CH_2)_3-$, arN, substd/cyc
amines            4    19    57

also $C_3H_5N_2$, 69.0451 (unsatd/ar amines/imines/azo);
$C_2HN_2O$, 69.0087; $C_2H_3N_3$, 69.0325 ($arN_3$); $C_3HS$,
68.9802 (arS); $CHN_4$, 69.0198 ($arN_4$)

| m/z 70 (37%) | | 8% | 28% |
|---|---|---|---|

$\underline{C_5H_{10}}$, 70.0782 $H-C_5H_{10}-Y*$, $H-C_5H_{10}-R-Y*$,
$H-R-CH(CH_3)C(CH_3)=CH_2$ etc, hc    25    27    82

$\underline{C_4H_6O}$, 70.0418 cyc/substd ketones/-O-/-OH 21    27    59

$\underline{C_4H_8N}$, 70.0656 cyc/subst amines (pyrro-
lidinyl, $CH_3N=CHCH_2CH_2-$, aziridinyl-$CH_2-$,
cyc-$CH_2CH(CH_3)CH(NH_2)-$, etc)    12    30    58

$\underline{C_3H_2O_2}$, 70.0054 $-CH_2O-CO-CH_2-$,
$-CH_2CH(-)-CO-O-$, 4-pyrones,
$-CH_2CH_2O-CO-$, $-CO-CH_2-CO-$, HO-ar(C=O),
etc           6    25    43

$\underline{C_3H_4NO}$, 70.0292 $-CH_2CH_2-CO-NH-$,
cyc-$N(C_2H_5)-CO-$, $CH_3C(CN)(OH)-$,
$H_2N-CO-CH=CH-$, $NCO-CH(CH_3)-$, $OCN-C_2H_4-$,
$-CH(CH_3)-CO-NH-$, etc      4    27    44

36

| m/z, comp | Substructure, neighbor | Prop | Abnd | Spcf |
|---|---|---|---|---|

$C_3H_6N_2$, 70.0529  $C_3H_7$-N=N-, ar$N_2$, cyc$N_2$,
etc                                                   3    28    46
also $C_2H_2N_2O$, 70.0165 (HO-ar$N_2$); $C_2NO_2$, 69.9928
(-CO-NH-CO-); $C_2H_4N_3$, 70.0403

| m/z 71 (38%) | | 10% | 33% |
|---|---|---|---|

$C_4H_7O$, 71.0496                                     29    35    62
-$(CH_2)_3$-CO-: $CH_2$ 70%; -O- 25%, $CH_2$ 20%  15+   32    50
$C_3H_7$-CO-: $CH_2$ 35%; -O- 25%                  7    60    60
-$(CH_2)_4$O-, tetrahydrofuryl-, 1,2-epoxy-
butyl-, -$(CH_2)_3$CH(OH)-, -$(CH_2)_3$OCH_2-,
-$CH_2CH_2$CH($CH_2$OH)-, -$CH_2CH_2OCH_2CH_2$-   22    30    75

$C_5H_{11}$, 71.0860                                   20    35    84

$C_3H_3O_2$, 71.0133                                   11    30    49
-$CH_2CH_2$-CO-O-, -$CH_2$-CO-OCH_2-               30    35    50
-$CH_2CH_2$O-CO-, -CH($CH_3$)-CO-O-,
-CO-$CH_2$-CO-, cyc-$CH_2$CH(O-)CH(O-)-           30    30    45

$C_4H_9N$, 71.0734 -$(CH_2)_4$N(-)-,
($CH_3)_2$NCH_2CH_2-, arN-$C_3H_7$, $H_2$N$(CH_2)_4$-,
cycN                                                   6    22    61

$C_3H_5NO$, 71.0370                                    3    21    40
also $C_3H_3S$, 70.9959; $C_3H_7N_2$, 71.0607; $C_2H_3N_2O$, 71.0244;
$C_2HNO_2$, 71.0006; $C_2N_2F$, 71.0044

| m/z, 72 (26%) | | 8% | 19% |
|---|---|---|---|

$C_4H_8O$, 72.0575                                     22    16    63
$C_2H_5$-CO-CH_2-, $CH_3$-CO-CH($CH_3$)-: $CH_2$ 75%  7    50+   85
HO-cyc, $CH_3$O-cyc, $C_2H_5$OCH_2CH_2-,
$CH_3$OCH_2CH_2CH_2-                                  10    25    50

| m/z, comp | Substructure, neighbor | Prop | Abnd | Spcf |
|---|---|---|---|---|
| $\underline{C_3H_4O_2}$, 72.0211 $C_2H_5O-CO-$ , $CH_3O-CO-CH_2-$ | | 13 | 15 | 56 |
| $\underline{C_3H_6NO}$, 72.0448 | | 8 | 33 | 60 |
| $(CH_3)_2N-CO-$ : $-NH-$ , ar, $-N(-)-$ | | 16 | 75 | 81 |
| $CH_3-CO-NHCH_2-$ : $CH_2$ 70% | | 18 | 25 | 80 |
| $C_2H_5NH-CO-$ : $-NH-$ , C | | 3 | 53 | 56 |
| $H_2N-CO-CH_2CH_2-$ : $CH_2$ , CH | | 3 | 33 | 50 |
| also $HON=CHCH_2CH_2-$ | | | | |
| $\underline{C_4H_{10}N}$, 72.0812 | | 10 | 26 | 64 |
| $C_3H_7NHCH_2-$ , $C_3H_7CH(NH_2)-$ , | | | | |
| $(CH_3)_2NCH(CH_3)-$ , etc | | 20 | 65 | 65 |
| $-CH(C_3H_7)NH-$ , $-CH_2N(C_3H_7)-$ , etc | | 5 | 85 | 90 |
| $(C_2H_5)_2N-$ , $C_3H_7NHCH(-)-$ , $C_4H_9NH-$ , | | | | |
| $C_4H_9N(-)-$ | | 10 | 33 | 55 |
| $\underline{C_2H_2NO_2}$, 72.0084 $-CH(-CO-OH)NH-$ , | | | | |
| $-N(CH_3)-CO-O-$ , $-CH(NH_2)-CO-O-$ , | | | | |
| $-CH_2O-CO-NH$ | | 3 | 21 | 50 |
| $\underline{C_2H_4N_2O}$, 72.0322 $-N(CH_3)-CO-N(-)-$ | | 2 | 24 | 55 |
| also $C_3H_8N_2$, 72.0686; $C_2H_2NS$, 71.9910 ($SCNCH_2-$); $C_3H_5P$, 72.0129 | | | | |

| m/z 73 (32%) | | 14% | 38+% | |
|---|---|---|---|---|
| $\underline{C_3H_5O_2}$, 73.0289 | | 21 | 30 | 67 |
| $-CH_2CH_2-CO-O-$ : $CH_2$ 60%; Si 50% | | 7 | 60+ | 46 |
| $HO-CO-CH_2CH_2-$ : $CH_2$ 55%, CH 16%, S 12% | | 7 | 48 | 66 |
| $C_2H_5O-CO-$ : $CH_2$ 55%, CH 18% | | 12 | 19 | 73 |
| $CH_3O-CO-CH_2-$ (also m/z 74): $CH_2$ 37%, | | | | |
| CH 34% | | 5 | 26 | 41 |
| $CH_3-CO-OCH_2-$ : $CH_2$ 50%, CH 30% | | 7 | 20+ | 65 |
| $-CH_2OCH_2CH_2O-$ , cyc-$CH_2CH_2OCH(-)-O-$ , | | | | |
| $-CH_2CH(OH)CH(OH)-$ | | 10 | 45 | 75 |

| m/z, comp | Substructure, neighbor | Prop | Abnd | Spcf |
|---|---|---|---|---|
| $\underline{C_4H_9O}$, 73.0653 | | 12 | 34 | 63 |
| | $-(CH_2)_4O-$ : $CH_2$ 75%; Si 50% | 13+ | 34 | 57 |
| | $C_2H_5CH(OCH_3)-$ : $CH_2$ 80% | 5 | 80+ | 90 |
| | $CH_3(CH_2)_3O-$ : C=O 80% | 6 | 19 | 40 |
| | $C_2H_5-CO-$ (also m/z 72): $CH_2$ 80% | 5 | 20+ | 94 |
| | $C_2H_5CH_2CH(OH)-$, $(CH_3)_2CH(OH)-$, | | | |
| | $-(CH_2)_3OCH_2-$, $(CH_3)_2CHOCH_2-$, | | | |
| | $C_2H_5CH_2OCH_2-$, $C_2H_5C(CH_3)(OH)-$, | | | |
| | $cyc-C(CH_3)(C_2H_5)O-$ | 20 | 40 | 65 |
| $\underline{C_3H_7NO}$, 73.0526 | | 4 | 35 | 70 |
| | $CH_3-CO-NH-CH_2-$ (also m/z 72): $CH_2$ 85% | 19 | 34 | 85 |
| | $-CH_2CH=NOCH_3$, $CH_3C(=NOCH_3)-$ | | | |
| | $-CH_2CH(=NOCH_3)-$ : $CH_2$; C=O | 9 | 80 | 99 |
| | $-CH_2CH_2C(=N-OH)-$ : $CH_2$ 50%; $CH_2$ 50% | 5 | 75 | 99 |
| | $CH_3-CO-N(CH_3)-$ | 3 | 55- | 55 |
| | $C_2H_5-CO-NH-$ | 2 | 30 | 99 |
| | $(CH_3)_2C=N-O-$, $-CH(OCH_3)-CH(-NH-)-$, | | | |
| | $-CH_2-CO-NH-CH_2-$, $CH_3NHCH_2CH(-)-O-$ | 10 | 80 | 80 |
| $\underline{C_2H_3NO_2}$, 73.0163 | | 3 | 35 | 68 |
| | $-NHCH_2-CO-O-$ : C=O; Si | 13 | 80 | 76 |
| | $-NHCH(-)-CO-O-$, $-N(-)-CH_2-CO-O-$ | 15+ | 50+ | 80 |
| | $HO-CO-CH(NH_2)-$ : $CH_2$ | 10 | 14 | 80 |
| | $ar-CH=C(NO_2)-$ | 5 | 10 | 59 |
| $\underline{C_2HO_3}$, 72.9925 | | 2 | 28 | 70 |
| | $-OCH_2-CO-O-$ : $CH_2$, ar, Si; Si, $CH_2$ | 10 | 73 | 53 |
| | $-OCH(-)-CO-O-$, $-OC(-)_2-CO-O-$, | | | |
| | $HOCH(-)-CO-O-$ | 22 | 55+ | 50 |
| | $-CO-CO-O-$ : ar, $-O-$; Si, $CH_2$ | 5 | 50- | 60 |
| $\underline{C_3H_9Si}$, 73.0373 $(CH_3)_3Si-$, $-CH_2Si(CH_3)_2-$ | | 1 | 44 | 80 |
| $\underline{C_2H_5N_2O}$, 73.0400 | | 1 | 23 | 70 |

MASS SPECTRAL CORRELATIONS

| m/z, comp    Substructure, neighbor | Prop | Abnd | Spcf |
|---|---|---|---|
| O=N-N($C_2H_5$)-: $CH_2$ | 16- | 45- | 80 |
| also $C_3H_9N_2$, 73.0764 (($CH_3$)$_2$N-N($CH_3$)-) | | | |

| m/z 74 (34%) | 5% | 27% | |
|---|---|---|---|

| $C_3H_6O_2$, 74.0367 | 30 | 43 | 84 |
|---|---|---|---|
| $CH_3$O-CO-$CH_2$-: $CH_2$ 70%, CH 13% | 51+ | 63 | 84 |
| $CH_3$O-CO-CH=: =CH- | 3 | 31 | 95 |
| HO-CO-CH($CH_3$)-: $CH_2$ 70%, -NH- 20% | 2 | 60 | 75 |
| cyc-CH(OH)CH(O$CH_3$)- | 2 | 50 | 45 |

| $C_6H_2$, 74.0156 | 23 | 13 | 92 |
|---|---|---|---|

| $C_2H_4NO_2$, 74.0241 | 5 | 20 | 81 |
|---|---|---|---|
| HO-CO-CH($NH_2$)-: $CH_2$ 75%, CH 20% | 30 | 40 | 77 |
| -$CH_2$O-CO-N(-)- | 2 | 70 | 58 |

| $C_4H_{10}$O, 74.0731 $CH_3$O-cyc, cycO, HO-cyc | 5 | 18 | 69 |
|---|---|---|---|

| $C_2H_2O_3$, 74.0003 cyc-OCH(O-)CH(OH)-, | | | |
|---|---|---|---|
| -OC(O$CH_3$)O-, -OC(-O$CH_2$-)O- | 3 | 23 | 55 |

| $C_3H_6S$, 74.0193 cyc-SC($CH_3$)$_2$-, -($CH_2$)$_3$S-, | | | |
|---|---|---|---|
| -$CH_2CH_2$S$CH_2$- | 3 | 39 | 64 |

| $C_3H_8NO$, 74.0605 -CH(OH)CH(NHC$H_3$)-, | | | |
|---|---|---|---|
| -CH(OH)CH($CH_3$)NH-, $CH_3$CH(OH)CH($NH_2$)- | 2 | 46 | 60 |

also $C_5$N, 74.0030 (arN); $C_3F_2$, 73.9968 ($F_2$-ar);
  $CH_2N_2O_2$, 74.0114 (ONNH-CO-); $C_2H_4$NS, 74.0067 (arNS);
  $CH_2N_2$S, 73.9940 ($H_2$N-arNS, -S-ar$N_2$); $C_3H_3$Cl, 73.9923
  (Cl-ar)

| m/z 75 (36%) | 8% | 30% | |
|---|---|---|---|

| $C_6H_3$, 75.0235 | 30 | 20 | 88 |
|---|---|---|---|

| $C_3H_7O_2$, 75.0445 | 13 | 30 | 80 |
|---|---|---|---|

| m/z, comp | Substructure, neighbor | Prop | Abnd | Spcf |
|---|---|---|---|---|
| $CH_3O-CO-CH_2-$: $CH_2$ 65%, CH 15% | | 24+ | 22 | 93 |
| $(CH_3O)_2CH-$: $CH_2$ 70%, CH 20% | | 6 | 75- | 80 |
| $C_2H_5-CO-O$: $CH_2$ 70%, CH 20% | | 6 | 29 | 74 |
| $HOC_2H_4OCH_2-$, $HOC_2H_4CH(OH)-$, $CH_3OCH_2CH(OH)$ etc | | 9 | 22 | 85 |
| $CH_3-CO-OCH_2-$ also $C_2H_5OCH(-)O-$ | | 2 | 42 | 70 |

| $\underline{C_5}HN$, 75.0108 arN | | 9 | 19 | 74 |
|---|---|---|---|---|

| $\underline{C_2H_7}OSi$, 75.0165 $-Si(CH_3)_2-O-$, $(CH_3)_2Si(OH)-$ | | 3 | 75 | 96 |
|---|---|---|---|---|

| $\underline{C_3H_4}Cl$, 75.0001 | | 4 | 31 | 71 |
|---|---|---|---|---|

| $\underline{C_3H_7}S$, 75.0271 | | 3 | 42 | 66 |
|---|---|---|---|---|
| $C_2H_5SCH_2-$, $CH_3SCH(CH_3)-$ also $C_3H_7S-$ | | 20 | 95 | 55 |

| $\underline{C_2H_5}NO_2$, 75.0319 $HO-CO-CH(NH_2)-$, $-NHCH_2-CO-O-$, $NH_2-CO-OCH_2-$ | | 3 | 28 | 78 |
|---|---|---|---|---|

| $\underline{C_2H_3}O_3$, 75.0082 $cyc-CH(-O-)-CO-O-$, $-OCH-CO-O-$ | | 2 | 27 | 55 |
|---|---|---|---|---|

also $C_2H_3OS$, 74.9908; $C_3HF_2$, 75.0046 ($ar-F_2$); $C_3H_9NO$, 75.0683 ($CH_3CH(OH)CH(NH_2)-$); $C_2OCl$, 74.9637; $C_2H_7N_2O$, 75.0556 ($CH_3-CO-NH-NH-$); $C_4H_8F$, 75.0610

| m/z 76 (30%) | | 10% | 13% | |
|---|---|---|---|---|

| $\underline{C_6H_4}$, 76.0313 | | 38 | 10 | 86 |
|---|---|---|---|---|

| $\underline{C_5H_2}N$, 76.0186 arN | | 10 | 13 | 61 |
|---|---|---|---|---|

| $\underline{C_5}O$, 75.9949 ar(CO), arO | | 6 | 21 | 63 |
|---|---|---|---|---|

| m/z, comp | Substructure, neighbor | Prop | Abnd | Spcf |
|---|---|---|---|---|
| $\underline{C_4N_2}$, 76.0060 $arN_2$ | | 4 | 16 | 49 |
| $\underline{C_3H_5Cl}$, 76.0079 $ClC_3H_6-$ | | 2 | 19 | 82 |
| $\underline{C_3H_8S}$, 76.0350 $C_3H_7S-$ | | 2 | 12 | 72 |
| $\underline{CS_2}$, 75.9448 $arS_2$, $-S-C(=S)-$, pyrolysis product | | 1 | 21 | 90 |
| $\underline{C_2H_4O_3}$, 76.0160 $CH_3O-CO-O-$, $HO-CO-CH(OH)-$, $-CH(OH)-CO-O$ | | 1 | 12 | 55 |
| also $C_3H_2F_2$, 76.0124; $H_2N_3O_2$, 76.0144 | | | | |
| m/z 77 (55%) | | 18% | 29% | |
| $\underline{C_6H_5}$, 77.0391 phenyl-Y, Y-phenyl-Y' | | 41 | 27 | 86 |
| $\underline{C_5H_3N}$, 77.0265 arN, ar-N(-)- | | 10 | 30 | 56 |
| $\underline{C_5HO}$, 77.0027 ar-CO-, etc | | 6 | 26 | 60 |
| $\underline{C_4HN_2}$, 77.0138 $arN_2$, arN-N(-)-, etc | | 2 | 26 | 40 |
| $\underline{C_3H_3F_2}$, 77.0203 $-C_2H_4CF_2-$, $CF_2=CHCH_2-$, etc | | 1 | 57 | 80 |
| $\underline{C_3H_6Cl}$, 77.0157 $C_2H_5CHCl-$, etc | | 1 | 26 | 71 |
| $\underline{CH_2O_2P}$, 76.9792 $CH_3OP(=O)(-)-$, etc | | 1 | 24 | 70 |
| m/z 78 (39%) | | 7% | 18% | |
| $\underline{C_6H_6}$, 78.0469 phenyl | | 42 | 15 | 83 |
| $\underline{C_5H_4N}$, 78.0343 pyridyl-, other arN, ar-NH- | | 11 | 21 | 51 |

| m/z, comp | Substructure, neighbor | Prop | Abnd | Spcf |
|---|---|---|---|---|
| $\underline{C_5H_2O}$, 78.0105 ar-CO-, arO | | 7 | 22 | 56 |
| $\underline{C_4H_2N_2}$, 78.0216 arN$_2$, ar-N=N- | | 5 | 20 | 43 |
| $\underline{C_4NO}$, 77.9979 arN-CO-, ar-NO, arNO, etc | | 2 | 19 | 36 |
| $\underline{CH_2S_2}$, 77.9604 -CH$_2$-S-S-, -SCH$_2$S-, SCH(-)S- | | 1 | 44 | 88 |

also $C_3N_3$, 78.0090; $C_2OF_2$, 77.9917 (-CF$_2$-CO-)

| m/z 79 (43%) | | 11% | 30% | |
|---|---|---|---|---|
| $\underline{C_6H_7}$, 79.0547 | | 32 | 34 | 89 |
| $\underline{C_5H_3O}$, 79.0184 ar-O-, ar-CH$_2$OH, subst cyc(C=O), etc | | 14 | 27 | 66 |
| $\underline{C_5H_5N}$, 79.0421 arN, subst cyc-NH-, etc | | 10 | 21 | 55 |
| $\underline{C_4H_3N_2}$, 79.0295 arN$_2$, etc | | 3 | 20 | 38 |
| $\underline{C_4HNO}$, 79.0057 arN-CO-, arNO, ar-NO, etc | | 2 | 20 | 34 |
| $\underline{CH_4O_2P}$, 78.9949 CH$_3$OP(=O)(-)$_2$, CH$_3$P(=O)(-)O- | | 1 | 34 | 85 |
| $\underline{CH_3O_2S}$, 78.9857 CH$_3$OS(=O)-, -CH$_2$OS(=O)- | | 1 | 42 | 78 |
| Br, 78.9183 | | 2 | 14 | 93 |

also CH$_3$S$_2$, 78.9683 (CH$_3$-S-S-); C$_2$H$_4$OCl, 78.9950 (ClCH$_2$OCH$_2$-, ClCH$_2$CH(OH)-); C$_2$HFCl, 78.9750; O$_3$P, 78.9585 (-OPH(=O)O-, -OP(-)(=O)O-)

| m/z 80 (28%) | | 8% | 17% | |
|---|---|---|---|---|
| $\underline{C_6H_8}$, 80.0626 cyclohexenes, etc | | 29 | 18 | 91 |

# MASS SPECTRAL CORRELATIONS

| m/z, comp | Substructure, neighbor | Prop | Abnd | Spcf |
|---|---|---|---|---|
| $C_5H_4O$, 80.0262 arO, etc | | 13 | 16 | 61 |
| $C_5H_6N$, 80.0499 arN (pyridyl, pyrrolyl-$CH_2$-), ar-$NH_2$, subst cycloalkanones | | 9 | 15 | 57 |
| $C_4H_2NO$, 80.0135 arN-OH, ar(N-CO-), ar-NO, etc | | 4 | 15 | 44 |
| $C_4H_4N_2$, 80.0373 ar$N_2$, arN-$NH_2$ | | 4 | 14 | 44 |
| $C_3H_2N_3$, 80.0246 ar$N_3$, ar$N_2$-NH- | | 2 | 15 | 36 |
| $C_4O_2$, 79.9898 ar-COOH, -CO-C≡C-CO-, ar-CO-O- | | 2 | 14 | 42 |
| HBr, 79.9261 (can be impurity) | | 2 | 9 | 96 |
| $C_3N_2O$, 80.0009 ar$N_2$-O-, ar$N_2$(C=O), etc | | 1 | 12 | 42 |

also $CH_4O_2S$, 79.9935 ($CH_3OS(=O)$-); $HO_3P$, 79.9663 (-OPH(=O)O-); $CH_4S_2$, 79.9761 ($CH_3$-S-S-)

| m/z 81 (39%) | | | 6% | 36% |
|---|---|---|---|---|
| $C_6H_9$, 81.0704 polyisoprenes, polyunsatd/cyc hc | | 33 | 46 | 92 |
| $C_5H_5O$, 81.0340 | | 19 | 40 | 67 |
|    furyl-$CH_2$-: -O- 49%, -S- 28%, $CH_2$ 11% | | 3 | 94- | 55 |
|    unsatd/subst hc-CO- | | 25 | 40 | 75 |
|    ar-OH, subst/unsatd/cyc hc-OH | | 10 | 25 | 70 |
| $C_5H_7N$, 81.0577 arN ($CH_3$-pyrrolyl etc), unsatd/cyc/substd imine/amine, ar amine | | 6 | 23 | 60 |

| m/z, comp | Substructure, neighbor | Prop | Abnd | Spcf |
|---|---|---|---|---|

$\underline{C_4H_3NO}$, 81.0214 Nar(C=O), NarO, unsatd
  methoxime, unsatd oxyamine, etc     3   25   48

$\underline{C_4HO_2}$, 80.9976 subst/cyc-O-CO-, $arO_2$,
  -O-ar-O-, etc     3   13   47

also $C_4H_5N_2$, 81.0451 ($arN_2$, etc); $C_3HN_2O$, 81.0087;
$C_3H_3N_3$, 81.0325 ($arN_3$, etc); $C_2F_3$, 80.9952; $H_2O_3P$,
80.9741 (-OP(=O)O-, etc); $C_4HS$, 80.9802 (arS); $C_5H_2F$
81.0140 (ar-F); $C_2H_3OF_2$ ($CH_3OCF_2$-); $C_2H_3FCl$, 80.9907
(-$CH_2$CFCl-, -$CH_2$CHFCl)

| m/z 82 (33%) | | 7% | 29% |
|---|---|---|---|

$\underline{C_6H_{10}}$, 82.0782 cyclohexyl, subst/unsatd
  hc     30   32   84

$\underline{C_5H_6O}$, 82.0148 furyl-$CH_2$-, ketones,
  cyc-OH, cycO, etc     17   29   57

$\underline{C_5H_8N}$, 82.0656 pyrrolizidinyl,
  cyc/substd/unsatd amines/imines,
  NC-$C_4H_8$-     7   29   53

$\underline{C_4H_4NO}$, 82.0292 unsatd ketoamines
  ($CH_3$NHC(-)=CH-CO-), arNO
  ($CH_3$-isoxazolyl), arN(C=O), subst/cyc
  aminoethers, etc     4   25   46

$\underline{C_4H_2O_2}$, 82.0054 ar(C=O)$_2$, HO-CO-C≡CCH$_2$-,
  ar-CO-OH, unsatd-CO-O-     4   23   49

$\underline{C_4H_6N_2}$, 82.0529 $arN_2$ (pyrimidinyl-$CH_2$-),
  unsatd amines     2   27   49

also $C_3H_2N_2O$, 82.0165; $C_3H_4N_3$, 82.0403 (triazinyl);

m/z, comp      Substructure, neighbor      Prop Abnd Spcf

$C_3NO_2$, 81.9928; $C_4H_2S$, 81.9880; $CCl_2$, 81.9376;

$C_2H_2N_4$, 82.0276 ($arN_4$); $C_2HF_3$, 82.0030; $CF_2S$, 81.9692

| m/z 83 (38%) | | 10% | 34% |
|---|---|---|---|

$\underline{C_6H_{11}}$, 83.0860 cyclohexyl,
   $CH_3CH=CHC(CH_3)_2-$ etc        29   33   89

$\underline{C_5H_7O}$, 83.0496 cyc ketones
   ($-CH_2C(CH_3)_2-CO-$ etc), lactones,
   cyc/subst/unsatd $-O-/-OH/-CO-$
   (($CH_3)_2C=CH-CO-$)        22   29   69

$\underline{C_4H_3O_2}$, 83.0133 diketones, ketoesters,
   $ar-CO-OCH_3$, subst/cyc $-O-/-OH/-CO-$
   (HO-furyl-)        6   28   50

$\underline{C_5H_9N}$, 83.0734 $NC(CH_2)_4-$, cyc/substd
   amines (piperidyl)        4   29   57

$\underline{C_4H_5NO}$, 83.0370 unsatd/cyc
   $-NH-/-NH_2/C=O/-OH/-O-/-CH=N-OH$
   (cyc$-NH-CO-CH=C(CH_3)-$, $-CH=C(CH_3)NH-CO-$,
   $H_2N-CO-C(CH_3)=CH-$)        4   24   50

$\underline{CHCl_2}$, 82.9454 $CHCl_2$, $-CCl_2-$        2   54   88

$\underline{C_4H_7N_2}$, 83.0607 $-CH=CHC(CH_3)=NNH-$ etc    2   37   57
also $C_3H_3N_2O$, 83.0244 ($arN_2(C=O)$, $arN_2-OH$, etc);
   $C_3HNO_2$, 83.0006; $C_3H_5N_3$, 83.0481; $C_5H_4F$, 83.0297;
   $C_4H_3S$, 82.9959 (thiophenyl-); $C_2H_2F_3$, 83.0108; $FO_2S$,
   82.9603 (F-S(=O)O-)

| m/z, comp | Substructure, neighbor | Prop | Abnd | Spcf |
|---|---|---|---|---|

$\underline{m/z}$ 84 (81%)                                                        12%    19%

$\underline{C}_5\underline{H}_8O$, 84.0575 cyc/subst ketones
(2-R-cyclopentanones), cyc/substd/unsatd
-O-/-OH                                                           21      18     57

$\underline{C}_6\underline{H}_{12}$, 84.0938 H-$C_6H_{12}$-Y*, H-$C_6H_{12}$-R-Y*,
etc                                                               23      14     77

$\underline{C}_4\underline{H}_4O_2$, 84.0211 beta diketones,
$CH_3$-4-pyrones, substd/cyc -O-CO-,
HO-CO-                                                            10      19     50

$\underline{C}_5\underline{H}_{10}N$, 84.0812 cyc/subst amines
(2-piperidynl, $\underline{N}$-$CH_3$-pyrrolidinyl,
cyc-$CH_2CH_2CH(NHC_2H_5)$- etc)                                   7      25     54

$\underline{C}_4\underline{H}_6NO$, 84.0448 subst/cyc amides
(-$CH_2CH(NH$-CO-$CH_3$, -$CH_2CH_2CH_2$-CO-NH-),
-$CH_2CH_2CH(NH_2)$-CO-, -CH($NH_2)CH_2CH_2$-CO-,
OCN-$C_3H_6$, etc                                                  6      22     45

$\underline{C}_4\underline{H}_8\underline{N}_2$, 84.0686 $C_4H_9$-N=N-,
-$(CH_2)_3C(-)$=N-NH-, -N(-)$(CH_2)_3N(-)$-,
ar$N_2$ etc                                                        2      28     49

$\underline{C}_3\underline{H}_4\underline{N}_2O$, 84.0322 HO-arN,
-CH(-)NH-CO-NHCH(-)-, $H_2$N-CO-arN,
$N_2$ar(C=O), etc                                                  2      23     48

$\underline{C}_3\underline{H}_2NO_2$, 84.0084 -N(-CO-$CH_3$)-CO-,
-C($NH_2$)=CH-CO-O-, HO-CO-arN, ar-$NO_2$           2      19     42
also $C_3O_3$, 83.9847 (-O-CO-$CH_2$-CO-)

| m/z, comp | Substructure, neighbor | Prop | Abnd | Spcf |
|---|---|---|---|---|
| m/z 85 (33%) | | 14% | 20% | |
| $\underline{C_5H_9O}$, 85.0653 | | 22 | 20 | 66 |
| | $-(CH_2)_4-CO-$: $CH_2$ 70%; $-O-$ 27% | 17+ | 25 | 63 |
| | $C_4H_9-CO-$: $CH_2$ 35%; $-CH(-)_2$ 30% | 6 | 40 | 65 |
| | cycO (tetrahydropyryl-), $-(CH_2)_5O-$, | | | |
| | $-(CH_2)_4CH(-)O-$, $-(CH_2)_3OCH_2CH_2-$, | | | |
| | $C_3H_7C(-)(CH_2OH)-$, $-(CH_2)_4OCH_2-$ | 16 | 32 | 65 |
| $\underline{C_6H_{13}}$, 85.1017 | | 19 | 17 | 85 |
| $\underline{C_4H_5O_2}$, 85.0289 | | 15 | 19 | 58 |
| | $CH_3-CO-CH_2-CO-$ | 4 | 60 | 82 |
| | gamma-lactones | 2 | 90 | 81 |
| | $-(CH_2)_3-CO-O-$, $-CH_2CH_2-CO-OCH_2-$, | | | |
| | $-CH_2CH_2O-CO-CH_2-$, 2-(methylethylene | | | |
| | ketal)-, $-CH_2CH(O-)CH_2CH(O-)-$, | | | |
| | $-O(CH_2)_3-CO-$ | 20 | 20 | 60 |
| $\underline{C_5H_{11}N}$, 85.0890 $(CH_3)_2N(CH_2)_3-$, | | | | |
| | $-(CH_2)_5N(-)-$, $arN-C_4H_9$, $C_4H_9CH=N-$ | 4 | 12 | 63 |

also $C_4H_7NO$, 85.0526; $C_3H_3NO_2$, 85.0163; $C_3HO_3$, 84.9925;
CClF$_2$, 84.9656; $C_4H_9N_2$, 85.0764 (cycN$_2$,
$(CH_3)_2N-N=CH-CH_2-$); $C_4H_5S$, 85.0115 (cycS)

| m/z, comp | Substructure, neighbor | Prop | Abnd | Spcf |
|---|---|---|---|---|
| m/z 86 (22%) | | 6% | 18% | |
| $\underline{C_5H_{12}N}$, 86.0968 | | 8 | 36 | 62 |
| | $(C_2H_5)_2NCH_2-$, $C_4H_9NHCH_2-$, | | | |
| | $C_3H_7N(CH_3)CH_2-$, $C_4H_9CH(NH_2)-$, | | | |
| | $(CH_3)_2NC(CH_3)_2-$, etc | 15 | 75 | 70 |
| | $-CH_2N(C_4H_9)-$, $-CH(CH_3)N(CH_3)CH(CH_3)-$, | | | |
| | $-CH_2CH_2N(C_2H_5)CH_2-$ | 8 | 55 | 60 |

| m/z, comp | Substructure, neighbor | Prop | Abnd | Spcf |
|---|---|---|---|---|
| $\underline{C_4H_6O_2}$, 86.0367 | 1-dioxolanyl(-)-cyc-CH$_2$- | 14 | 16 | 54 |

$\underline{C_5H_{10}O}$, 86.0731                 13   17   53
    C$_3$H$_7$-CO-CH$_2$-,   C$_2$H$_5$-CO-CH(CH$_3$)-,
      Y*-C$_5$H$_{10}$O-                       7   20   63

$\underline{C_4H_8NO}$, 86.0605   H$_2$N-CO-CH$_2$CH(CH$_3$)-,
    CH$_3$NH-cyc-OH                       8   24   57

$\underline{C_3H_4NO_2}$, 86.0241   -CH(-CO-OCH$_3$)-NH-,
    -CH(NH$_2$)-CO-OCH$_2$-                  4   27   44
also C$_3$H$_6$N$_2$O, 86.0478 (-NH-CO-NH-); C$_3$H$_2$O$_3$, 86.0003;
  C$_4$H$_{10}$N$_2$, 86.0842 ((CH$_3$)$_2$NN=CH-CH$_2$-); C$_2$H$_4$N$_3$O, 86.0351
  (H$_2$N-CO-NH-N=CH-)

| m/z 87 (23%) | | 7% | 21% |
|---|---|---|---|
| $\underline{C_4H_7O_2}$, 87.0445 | | 25 | 26 | 62 |

    CH$_3$O-CO-CH$_2$CH$_2$-: CH$_2$ 55%, -CH= 15%    12   34   35
    CH$_3$O-CO-CH(-)-CH$_2$-: CH$_2$ 70%; CH$_2$ 55%    5   50   58
    CH$_3$O-CO-CH=CH-: CH$_2$                3   53-   99
    -O-CO(CH$_2$)$_3$-,   HO-CO-(CH$_2$)$_3$-,
     -CO-O(CH$_2$)$_3$-,   -CH$_2$O-CO-(CH$_2$)$_2$-,
     -CO-(CH$_2$)$_3$-O-,   CH$_3$-dioxolanyl-,
     CH$_3$CO-O(CH$_2$)$_2$-                  2   23   55

$\underline{C_5H_{11}O}$, 87.0809                 11   22   60
    HOC(C$_2$H$_5$)$_2$-,   HOC(C$_3$H$_7$)(CH$_3$)-,
     HOCH(C$_4$H$_9$)-                  11   50-   80
    C$_4$H$_9$OCH$_2$-,   C$_2$H$_5$OC(CH$_3$)$_2$-,
     CH$_3$OCH(C$_3$H$_7$)-                7   20-   70
    -(CH$_2$)$_5$-O-,   -(CH$_2$)$_4$OCH$_2$-,   -(CH$_2$)$_3$O(CH$_2$)$_2$-,
     -CH(OH)(CH$_2$)$_4$-,   -CH$_2$C(OH)(-)C$_3$H$_6$-    16   35   40

MASS SPECTRAL CORRELATIONS

| m/z, comp | Substructure, neighbor | Prop | Abnd | Spcf |
|---|---|---|---|---|

$C_4H_9NO$, 87.0683 — 4 25 74

$CH_3ON=C(CH_3)CH_2-$: $CH_2$ — 8 90 99

$(CH_3)_2N-CO-CH_2-$ — 2 99+ 99

$CH_3CO-NH(CH_2)_2-$, $C_3H_7CO-NH-$,

$CH_3CO-N(CH_3)-CH_2-$ — 28 18 85

$C_3H_3O_3$, 87.0082 $-CO-CH_2CO-O-$,

$-CO-CH(-)-CO-O-$, $-CO-O-CH_2CH_2O-$,

$-CH(O-)-CH(OH)-CH(O-)-$, $-CH(OH)CH_2O-CO-$,

$-CH(C_2H_5)O-CO-$, $-(CH)_3(-)(OH)(O-)-O-$ — 4 19 56

$C_4H_7S$, 87.0271 — 3 20 83

Thiacyclopentyl-: $CH_2$ — 5 99 75

other thiacycloalkanes (cycS) — 75 20 70

$C_3H_5NO_2$, 87.0319 — 3 20 62

$-CH_2C(NH_2)(COOH)-$ — 7 37- 99

$CH_3O-CO-CH(NH_2)-$ — 3 59- 99

$CH_3CO-N(-)-CO-$, $-CH(OH)CH_2N(-)-CO-$,

$-NHCH(-)-CO-OCH_2-$, $CH_3O-CO-CH_2-NH-$,

$(arN)-CO-OCH_2-$, $-C(=NOH)-C(OH)(-)-$ — 30 25 75

also $C_5H_{13}N$, 87.1047 ($(C_2H_5)_2NCH_2-$); $C_4H_4Cl$, 87.0001;
$C_3H_7N_2O$, 87.0556; $C_4H_{11}Si$, 87.0530 ($C_2H_5Si(CH_3)_2-$)

| m/z 88 (17%) | | 5% | 22% |
|---|---|---|---|

$C_4H_8O_2$, 88.0524 — 22 29 66

$CH_3O-CO-CH(CH_3)-$: $CH_2$ 70% — 13 80 92

$C_2H_5O-CO-CH_2-$: $CH_2$ 55%, C=O 15%, CH 15% 16 52 78

$HO-CO-CH(C_2H_5)-$, $HO-CO-C(CH_3)_2-$ — 4 30 75

cyc-$CH(OCH_3)CH(OCH_3)-$ — 2 56 22

also $C_2H_5OC(-)(CH_3)O-$

| m/z, comp | Substructure, neighbor | Prop | Abnd | Spcf |
|---|---|---|---|---|

C₇H₄, 88.0131 ext/substd ar  ...  16  10  84

$\underline{C_3H_6NO_2}$, 88.0397  ...  6  26  78
  -CH(-CO-OCH₃)-NH-: CH₂ 60%, C=O 70%  ...  31  32  80
  CH₃O-CO-CH₂NH-: C=O 65%, Si, 15%  ...  9  25  80
  also CH₃O-CO-CH(NH₂)-

$\underline{C_3H_4O_3}$, 88.0160  ...  4  24  48
  -CH(O-)CH(OCH₃)O-  ...  9  67  33
also C₄H₅Cl, 88.0079; C₅H₁₂O, 88.0887 (CH₃O-cyc);
  C₆H₂N, 88.0186 (arN); C₄H₁₀NO, 88.0661
  (CH₃OCH₂NHCH(CH₃)-); C₄H₈S, 88.0350
  (-CH(CH₃)CH₂SCH₂-); C₃H₈N₂O, 88.0635

m/z 89 (28%)  ...  11%  15%

$\underline{C_7H_5}$, 89.0391 ext-arY, unsatd-ar  ...  36  12  83

$\underline{C_6H_3N}$, 89.0265 arN  ...  9  12  61

$\underline{C_4H_9O_2}$, 89.0602  ...  7  17  78
  C₃H₇-CO-O-: CH₂ 85%  ...  10  32  74
  also C₃H₇OCH(-)O-

$\underline{C_3H_5O_3}$, 89.0238 CH₃-CO-CO-O-  ...  3  25  67

$\underline{C_6HO}$, 89.0027 arO, ar-CO-  ...  5  13  65

$\underline{C_4H_9S}$, 89.0428 C₃H₇SCH₂-, C₂H₅SCH(CH₃)-  ...  2  28  61

$\underline{C_3H_7NO_2}$, 89.0475 HO-CO-CH(NH₂)CH₂-,
  CH₃O-CO-CH₂NH-, HO-CO-C(CH₃)(NH₂)-  ...  2  14  76

$\underline{C_3H_9OSi}$, 89.0322 -CH₂OSi(CH₃)₂-  ...  1  25  99
also C₃H₅OS, 89.0064 (1,3-oxathiolanyl); C₄H₆Cl,

| m/z, comp | Substructure, neighbor | Prop | Abnd | Spcf |
|---|---|---|---|---|
| 89.0157 ($ClCH=C(CH_3)CH_2-$); $C_5HN_2$, 89.0138 ($arN_2$) | | | | |

| m/z 90 (19%) | | 7% | 14% | |
|---|---|---|---|---|

| $C_7H_6$, 90.0469 | | 33 | 12 | 83 |
|---|---|---|---|---|

| $C_6H_4N$, 90.0343 arN, ar-NH- | | 14 | 11 | 62 |
|---|---|---|---|---|

| $C_6H_2O$, 90.0105 ar(CO), arO, ar-O- | | 7 | 13 | 56 |
|---|---|---|---|---|

| $C_5H_2N_2$, 90.0216 $arN_2$ | | 4 | 16 | 43 |
|---|---|---|---|---|

| $C_3H_6O_3$, 90.0316 | | 2 | 23 | 76 |
|---|---|---|---|---|
| $CH_3O-CO-CH(OH)-$: CH, $CH_2$ | | 25 | 70 | 87 |
| also $C_2H_5O-CO-O-$ | | | | |

| $C_3H_8NO_2$, 90.0554 $CH_3O-CO-CH_2NH-$ | | 1 | 22 | 80 |
|---|---|---|---|---|

also $C_4H_7Cl$, 90.0235 ($C_4H_8Cl-$); $C_4H_{10}S$, 90.0506
($C_4H_9S-$); $C_5NO$, 89.9979; $C_4H_{10}O_2$, 90.0680; $C_3H_6OS$,
90.0142; $C_4N_3$, 90.0090 ($arN_2-N(-)-$, $arN_3$)

| m/z 91 (46%) | | 11% | 38% | |
|---|---|---|---|---|

| $C_7H_7$, 91.0547 phenyl-$CH_2$-Y, phenyl-$C(-Y)_n$, | | | | |
|---|---|---|---|---|
| $CH_3$-phenyl-$Y_n$, etc | | 39 | 44 | 82 |

| $C_6H_5N$, 91.0421 cycN-ar, ar-NH-, arN | | 10 | 30 | 58 |
|---|---|---|---|---|

| $C_6H_3O$, 91.0184 ar-CO-, arO, etc | | 10 | 28 | 56 |
|---|---|---|---|---|

| $C_5H_3N_2$, 91.0295 $arN_2$, ar-N=N-, etc | | 2 | 36 | 40 |
|---|---|---|---|---|

| $C_4H_8Cl$, 91.0314 | | 1 | 46 | 70 |
|---|---|---|---|---|
| $Cl(CH_2)_4-$: $n-C_nH_{2n+1}$ | | 40 | 40 | 65 |
| $C_2H_5CCl(CH_3)-$ | | 8 | 72 | 75 |

| m/z, comp | Substructure, neighbor | Prop | Abnd | Spcf |
|---|---|---|---|---|

$C_5HNO$, 91.0057 ar-NO, cycN(C=O),
  arN(C=O), etc        1   25   37
also $C_3H_4OCl$, 90.9950 ($ClC_2H_4$-CO-); $C_2H_5NOS$, 91.0094;
  $CHNS_2$, 90.9556 (-C(=S)NH-); $C_3H_7O_3$, 91.0394
  ($C_2H_5OC(-)(-O-)_2$)

m/z 92 (31%)        11%   15%

$C_7H_8$, 92.0626 phenyl-$CH_2$- etc    32   13   85

$C_6H_4O$, 92.0262 arO, ar(C=O), ar-O-, ar-OH   12   13   55

$C_6H_6N$, 92.0499 pyridyl-$CH_2$-, ar-$NH_2$,
  ar-NH-        8   14   52

$C_5H_4N_2$, 92.0373 ar$N_2$, etc      4   16   43

$C_5H_2NO$, 92.0135 ar-NO, arN-CO-, etc   3   12   38
also $C_4H_2N_3$, 92.0246; $C_4N_2O$, 92.0009; $C_5O_2$, 91.9898;
  CHBr, 91.9261; $C_3H_8O_3$, 92.0473 ($HOCH_2CH(OH)CH(OH)$-)

m/z 93 (35%)        8%   34%

$C_7H_9$, 93.0704 terpenes, cyclohexenyl-,
  polyunsatd cyc hc     24   41   89

$C_6H_5O$, 93.0340 phenyl-O-, HO-phenyl-
  ar-CO-, ar-O-, subst/cyc(C=O), etc   15   26   66

$C_6H_7N$, 93.0577 pyridyl-$CH_2$-, phenyl-NH-,
  ar-amines, cyc-NH-, $CH_3$-pyridyl-   8   35   62

$C_5H_5N_2$, 93.0451 $CH_3$-pyrazinyl-,
  $R_2$N-pyridyl-, ar$N_2$, ar-N=N-, arN-$NH_2$   3   26   46

MASS SPECTRAL CORRELATIONS

| m/z, comp | Substructure, neighbor | Prop | Abnd | Spcf |
|---|---|---|---|---|
| $C_5H_3NO$, 93.0214 imidazoyl, arN-CO-, | | | | |
| cyclopentadienyl-NO, ar-NO, arN-OH | | 3 | 22 | 45 |
| $C_5HO_2$, 92.9976 ar-$(O-)_2$, arO-CO-, etc | | 3 | 27 | 42 |
| $C_4H_3N_3$, 93.0325 $arN_3$, etc | | 2 | 22 | 37 |
| $C_4HN_2O$, 93.0078 $arN_2(C=O)$, ar-NH-CO-NH- | | 2 | 36 | 41 |
| $C_3F_3$, 92.9952 unsatd, perhalocarbon | | 2 | 22 | 73 |
| $C_3H_6OCl$, 93.0106 $ClC_2H_4OCH_2-$, | | | | |
| $ClCH_2C(OH)(CH_3)-$, etc | | 1 | 50 | 65 |
| $CH_2Br$, 92.9339 $BrCH_2-$, cyc-Br | | 1 | 24 | 84 |

also $CH_2O_3P$, 92.9741 $(CH_3OP(=O)(-)O-)$; $C_2H_6ClSi$,
93.9922 $((CH_3)_2SiCl-)$

| m/z 94 (27%) | | 8% | 19% | |
|---|---|---|---|---|
| $C_7H_{10}$, 94.0782 polyunsatd/cyc hc | | 22 | 17 | 91 |
| $C_6H_6O$, 94.0418 | | 19 | 22 | 65 |
| $C_6H_5O-$: $CH_2$ 55%, C=O 15% | | 14 | 56 | 72 |
| also ar-OH, ar-$OCH_3$, substd alkanones, cyc-CO-O-, etc | | | | |
| $C_6H_8N$, 94.0656 subst amines, imines, etc | | 7 | 19 | 57 |
| $C_5H_4NO$, 94.0292 pyrrolyl-CO-, HO-arN-, | | | | |
| etc | | 4 | 17 | 48 |
| $C_5H_2O_2$, 94.0054 ar(C=O)-CO-, subst | | | | |
| esters, etc | | 4 | 14 | 44 |
| $C_5H_6N_2$, 94.0529 $arN_2$ ($CH_3$-pyrazinyl-) | | | | |
| subst/unsatd $cycN_2$ | | 3 | 14 | 44 |

m/z, comp    Substructure, neighbor    Prop Abnd Spcf

also $C_4H_2N_2O$, 94.0165 ($C_3H_2N_2$-CO-); $C_4H_4N_3$, 94.0403
(ext-ar$N_3$); $C_3H_2N_4$, 94.0276 (ext-ar$N_4$); $C_2Cl_2$,
93.9376; $C_2H_6S_2$, 93.9917 ($C_2H_5$S-S-)

| m/z 95 (34%) | 17% | 26% | |
|---|---|---|---|
| $\underline{C_7H_{11}}$, 95.0860 polyunsatd/cyc hc | 26 | 25 | 92 |
| $\underline{C_6H_7O}$, 95.0496 ar or cyc/subst/unsatd oxygen cpds | 16 | 25 | 67 |
| $\underline{C_5H_3O_2}$, 95.0133 furyl-CO-, ar(-O-CO-), subst/cyc -CO-OCH$_3$, etc | 5 | 23 | 55 |
| $\underline{C_6H_9N}$, 95.0734 cyc imine, nitriles, cyc/unsatd amine | 3 | 18 | 56 |
| $\underline{C_5H_5NO}$, 95.0370 pyridyl-O-, ar(N-CO-), subst/cyc-NH-CO-, ar-NO | 3 | 16 | 48 |
| $\underline{C_5H_7N_2}$, 95.0607 arN$_2$, cyc imine | 2 | 24 | 51 |
| $\underline{C_4H_3N_2O}$, 95.0244 imidazole-CO-, arN$_2$-CO-, arN$_2$-OH, etc | 2 | 20 | 48 |
| $\underline{C_4HNO_2}$, 95.0006 arN(CO)$_2$, -CO-ar-NH-CO- | 1 | 14 | 42 |

also $C_4H_5N_3$, 95.0481 (arN$_2$-NH$_2$); $C_3HN_3O$, 95.0117;
$C_2HCl_2$, 94.9454; $C_3H_3N_4$, 95.0355; $CH_4O_3P$, 94.9898
(CH$_3$OP(=O)O-); $C_6H_4F$, 95.0297 (ar-F)

| m/z 96 (28%) | 9% | 19% | |
|---|---|---|---|
| $\underline{C_7H_{12}}$, 96.0938 substd/cyc/unsatd hc | 23 | 19 | 81 |
| $\underline{C_6H_8O}$, 96.0575 R-CO-, substd/cyc ketones, cyc-O-, etc | 19 | 18 | 60 |

| m/z, comp | Substructure, neighbor | Prop | Abnd | Spcf |
|---|---|---|---|---|

$\underline{C_6H_{10}N}$, 96.0812 quinolizidinyl,
  substd/cyc amines, $NC-C_5H_{10}^-$       7    22    58

$\underline{C_5H_4O_2}$, 96.0211 unsatd esters, etc    5    13    47
also $C_5H_6NO$, 96.0448; $C_5H_8N_2$, 96.0686 (arN-NH-, etc),
  $C_4H_2NO_2$, 96.0084 ($NHar(C=O)_2$, etc); $C_4H_4N_2O$, 96.0322
  ($arN_2$-OH etc); $C_4H_6N_3$, 96.0559 ($C_2H_5$-$arN_3$); $C_2H_2Cl_2$,
  95.9532; $C_3H_4N_4$, 96.0433

| m/z 97 (32%) | | 15% | 21% |
|---|---|---|---|

$\underline{C_7H_{13}}$, 97.1017 $CH_3$-cyclohexyl etc     23    21    88

$\underline{C_6H_9O}$, 97.0653 cyc ketones
  (cyclopentyl-CO-, cyc-$CH_2CH(C_3H_7)$-CO-
  etc), epoxycyclohexyl, $C_3H_7CH=CH-CO-$
  etc, cyc/subst/unsatd -O-/-OH     21    19    71

$\underline{C_5H_5O_2}$, 97.0289 cyc/unsatd/subst
  $C=O/(C=O)_2/-O-/(-O-)_2/-OC(-)_2O-$(cyclo-
  pentadione)     7    21    52

$\underline{C_6H_{11}N}$, 97.0890 $NC(CH_2)_5^-$,
  $CH_3$-pyrrolizidinyl-, subst/cyc/unsatd
  amines     4    18    62

$\underline{C_5H_7NO}$, 97.0526 $CH_3$-oxazoles,
  cyc/unsatd -N(-)-/-NH-/-N=/C=O/-OH/-O-    3    15    45
also $C_4H_3NO_2$, 97.0163 (-CH=CH-CO-N(-)-CO-, ar-$NO_2$);
  $C_4H_5N_2O$, 97.0400 (carbamyl); $C_4HO_3$, 96.9925; $C_5H_9N_2$,
  97.0764; $C_5H_5S$, 97.0115 (thiophenyl-$CH_2$-, ar-$SCH_3$),
  $C_2H_3Cl_2$, 96.9611; $C_2OF_3$, 96.9901 (-$C_2HF_3$-O-,
  $CF_3$-CO-)

| m/z, comp | Substructure, neighbor | Prop | Abnd | Spcf |
|---|---|---|---|---|
| m/z 98 (26%) | | 10% | 19% | |

$\underline{C_7H_{14}}$, 98.1094 H-$C_7H_{14}$-Y*, H-$C_7H_{14}$-R-Y*    22    12    79

$\underline{C_6H_{10}O}$, 98.0731 2-R-cyclohexanone,
   R-$(CH_2)_5$-CO-Y, $C_2H_5$CH=CH-CO-$CH_2$-,
   cyc/subst ketones/-O-/-OH, etc    18    20    60

$\underline{C_5H_6O_2}$, 98.0367 furyl-CH(OH)-,
   2,3-$(CH_3)_2$-4-pyrones, substd/cyc esters
   etc    11    19    53

$\underline{C_6H_{12}N}$, 98.0968 cyc/substd amines,
   (piperidine-$CH_2$-), $C_4H_9$-CH=N$CH_2$- etc    6    29    60

$\underline{C_5H_8NO}$, 98.0605 cyc amides (valerolactams
   etc), $(CH_3)_2$NCH=CH-CO-,
   -$(CH_2)_3$CH($NH_2$)-CO-, OCN-$C_4H_8$-    5    25    50
also $C_4H_4NO_2$, 98.0241 ($H_2$N-CO-CH=CH-CO-); $C_5H_{10}N_2$,
   98.0842 ($C_5H_{11}$-N=N-); $C_4H_2O_3$, 98.0003
   (HO-CO-CH=CH-CO-, HO-CO-ar-CO-); $C_4H_6N_2O$, 98.0478;
   $C_4H_4NS$, 98.0065 (thiazole-$CH_2$-)

| m/z 99 (26%) | | 9% | 17% | |

$\underline{C_6H_{11}O}$, 99.0809    15    15    62
   $C_5H_{11}$-CO-: $CH_2$ 20%, -CH= 18%, -NH- 16%    9    40    65
   -$(CH_2)_5$-CO-, -$(CH_2)_5$CH(OH)-,
   -$(CH_2)_4$CH(OH)CH(-)-, -$CH_2$CH($C_4H_9$)O-    13    20    60

$\underline{C_5H_7O_2}$, 99.0445 delta- and $CH_3$-gamma-
   lactones, -$(CH_2)_4$-CO-O-, $CH_3$-CO-$C_2H_4$-CO-,
   1-dioxolanyl(-)-cyc-$CH_2CH_2$-    14    18    60

| m/z, comp | Substructure, neighbor | Prop | Abnd | Spcf |
|---|---|---|---|---|

$\underline{C_7H_{15}}$, 99.1173 satd hc                                    19    10    87

also $C_5H_9NO$, 99.0319 $(OCN-(CH_2)_4-)$; $C_4H_5NO_2$, 99.0319

  $(arN-CO-OC_2H_5)$; $C_6H_{13}N$, 99.1047; $C_4H_3O_3$, 99.0082;

$C_5H_{11}N_2$, 99.0920 $(cycN_2)$; $C_5H_7S$, 99.0271 $(cycS)$;

$C_4H_5NS$, 99.0145 $(thiazole-CH_2-)$

| m/z 100 (18%) | | 6% | 18% |
|---|---|---|---|

$\underline{C_5H_{10}NO}$, 100.0761 $(C_2H_5)_2N-CO-$,

  $-(CH_2)_4-NH-CO-$, $CH_3O-N=C(CH_3)CH(CH_3)-$      8    24    62

$\underline{C_5H_8O_2}$, 100.0524 $CH_3O-CO-CH=CH-CH_2^-$,

  $(CH_3-CO-)_2CH-$                                  13    15    59

$\underline{C_6H_{14}N}$, 100.1125 $C_5H_{11}CH(NH_2)-$,

  $cyc-CH_2CH(CH_3)C(CH_3)_2-N(-)-$,

  $-CH_2CH(CH_3)-N(C_2H_5)-CH_2-$                      7    17    68

$\underline{C_6H_{12}O}$, 100.0887 $C_2H_5-CO-C(CH_3)_2-$,

  $C_4H_9O-ar$                                        9    11    64

$\underline{C_4H_6NO_2}$, 100.0397 $-CH_2C(=NOCH_3)-CO-$,

  $-CH=CHCH_2NH-CO-O-$                                5    16    54

$\underline{C_4H_4O_3}$, 100.0160 substd trioxane,

  $HO-CO-C(-)_2CH_2-CO-$                              5    15    50

$\underline{C_8H_4}$, 100.0313 ar/unsatd hc             7    10    71

$\underline{C_5H_{12}N_2}$, 100.0998

  $cyc-CH_2CH_2N(CH_2CH_2CH_2NH-)-$,

  $-CH_2CH_2N(-)(CH_2)_3NH-$                          3    20    59

also $C_4H_8N_2O$, 100.0635 $(-NH(CH_2)_3N(-)-CO-)$; $C_3H_4N_2O_2$,

  100.0033 $(HO-CO-CH(-)NH-CO-)$; $C_2F_4$, 99.9936

| m/z, comp | Substructure, neighbor | Prop | Abnd | Spcf |
|---|---|---|---|---|

<u>m/z 101 (25%)</u>        8%   22%

$\underline{C_5H_9O_2}$, 101.0602 $C_2H_5O-CO-CH_2CH_2-$,
$C_4H_9O-CO-$, cyc-$CH_2CH_2C(OCH_3)_2-$,
$C_2H_5$-dioxolanyl-, -$(CH_2)_3-CO-O-CH_2-$     <u>19    21    60</u>

$\underline{C_8H_5}$, 101.0391 -styrenyl-$Y_2$*, ext-ar    <u>12    11    75</u>

$\underline{C_6H_{13}O}$, 101.0966 $C_5H_{11}CH(OH)-$,
$C_4H_9C(CH_3)(OH)-$, $CH_3CH(-)(CH_2)_3C(-)(OH)-$,
cyc-$(CH_2)_3CH(-CH_2CH_2O-)-$     <u>8    13    68</u>

$\underline{C_4H_5O_3}$, 101.0238 $CH_3O-CO-CH_2-CO-$,
-$O(CH_2)_3-CO-O-$, -$CH_2-CO-OCH_2CH_2O-$,
$CH_3O-CO-CH(-)-CO-$,
cyc-$CH_2CH(-O-)CH(-O-)-O-$     <u>7    29    57</u>

$\underline{C_7H_3N}$, 101.0265 ext-arN, NC-phenyl-     <u>6    9    67</u>

$\underline{C_4H_7NO_2}$, 101.0475 $HO-CO-CH(NH_2)CH_2CH_2-Y$,
$CH_3O-CO-CH=CH-N(-)-$, $C_2H_5O-CO-arN$     <u>3    18    62</u>
also $C_5H_{11}NO$, 101.0839 ($CH_3-CO-NH(CH_2)_3-$,
$(CH_3)_2N(CH_2)_3O-$); $C_5H_9S$, 101.0428
($CH_3$-thiacyclopentyl-); $C_4H_5OS$, 101.0064
(-$CH_2CH_2S-CO-CH_2-$); $CFCl_2$, 100.9360; $C_4H_6OP$,
101.0156 (-phospholane-O-); $C_4H_7NS$, 101.0299
($SCN-(CH_2)_3-$); $C_3H_7N_3O$, 101.0586
($H_2N-CO-NH-N=CH-CH_2-$)

<u>m/z 102 (23%)</u>        7%   15%

$\underline{C_7H_{12}}$, 96.0938 phenyl-$C(-)=CH-$, ext-ar,
quinolines     <u>23    12    86</u>

MASS SPECTRAL CORRELATIONS

| m/z, comp | Substructure, neighbor | Prop | Abnd | Spcf |
|---|---|---|---|---|
| $C_7H_4N$, 102.0343 ext-arN, phenyl-arN | | 11 | 12 | 61 |
| $C_5H_{10}O_2$, 102.0680 $C_2H_5O-CO-CH(CH_3)-$, $C_3H_7-CO-CH_2-$, $-(CH_2)_4-CO-O-$, $C_3H_7O-CO-CH(-)-$ | | 9 | 15 | 69 |
| $C_7H_2O$, 102.0105 ext-ar(C=O), ext-arO | | 8 | 12 | 71 |
| $C_4H_8NO_2$, 102.0554 $C_2H_5O-CO-CH(NH_2)-$, $C_2H_5O-CO-NHCH_2-$, $(CH_3-CO-)_2N-$, $-C_4H_8-ONO$, $CH_3O-CO-C(CH_3)(NH_2)-$, $CH_3-CO-NHCH_2CH(OH)-$, $-NH(CH_2)_3-CO-O-$ | | 4 | 22 | 65 |
| $C_6H_2N_2$, 102.0216 ext-arN$_2$, NC-pyridyl- | | 3 | 14 | 43 |
| $C_4H_6O_3$, 102.0316 $-CH(OH)CH(OH)CH(OH)CH(-)-$, $CH_3O-CO-CH(-CH_2OH)-$, $CH_3-CO-CH(-)-CO-O-$ | | 3 | 11 | 50 |

also $C_5H_{12}NO$, 102.0917 ($C_2H_5OCH_2CH_2CH(NH_2)-$);
  $C_3H_4NO_3$, 102.0190 (cyc-CH(OH)-CH(-NH-CO-)-)

| m/z 103 (32%) | | 12% | 17% | |
|---|---|---|---|---|
| $C_8H_7$, 103.0547 phenyl-$C_2H_2-$, indoles, benzofurans, phenyl-CH(-)-CH(-)- | | 29 | 13 | 85 |
| $C_7H_5N$, 103.0421 phenyl-arN, phenyl-CH=N-, ext-arN, cyc-CH(phenyl)-N(-)- | | 10 | 14 | 58 |
| $C_7H_3O$, 103.0184 ext-ar(C=O), Y*-phenyl-CO-Y*, ar-CH=CH-CO-, ext-ar-O- | | 7 | 15 | 64 |
| $C_5H_{11}O_2$, 103.0758 $(C_2H_5O)_2CH-$, $C_4H_9OCH(-)O-$, $C_4H_9-CO-O-$, $HOCH(CH_3)CH_2OCH(CH_3)-$, | | | | |

| m/z, comp | Substructure, neighbor | Prop | Abnd | Spcf |
|---|---|---|---|---|

$CH_3OCH_2CH_2CH(OCH_3)-$,  $C_4H_9O-CO-$        4    25    68

$\underline{C_4H_7O_3}$, 103.0394  $CH_3O-CO-CH_2CH(OH)-$,
  $CH_3-CO-CH_2-CO-O-$,  $CH_3O-CO-CH(OCH_3)-$,
  $cyc-CH_2-CH(CH_2OH)-OCH(-)-O-$        4    16    64

$\underline{C_4H_{11}OSi}$, 103.0479  $(CH_3)_3SiOCH_2-$,
  $-CH_2OCH_2-Si(CH_3)_2-$        2    27    90

$\underline{C_6H_3N_2}$, 103.0295  $ext-arN_2$,
  $cyc-NN(phenyl)-$,  $-NH-phenyl-N(-)-$        2    13    42

$\underline{C_5H_{11}S}$, 103.0584  $C_4H_9SCH_2-$,
  $C_2H_5SC(CH_3)_2-$,  $-CH(C_2H_5)SCH(CH_3)-$        1    24    55
also $C_5H_8Cl$, 103.0314
  ($cyc-CCl(-)-C(-)(CH_3)-C(-)(CH_3)-$);  $C_3H_3O_4$, 103.0031
  ($cyc-CH(OH)-CH(-CO-OH)-O-$)

| m/z 104 (26%) | | 10% | 19% |
|---|---|---|---|

$\underline{C_8H_8}$, 104.0626  tetralins, indans,
  $phenyl-CH_2CH_2-Y^*$,  $\underline{o}-CH_3-phenyl-CH_2-Y^*$,
  phenyl-cycR        27    20    84

$\underline{C_7H_6N}$, 104.0499  phenyl-arN, ext-arN,
  $cyc-CH(phenyl)-NH-$,  $phenyl-C(-)=N-$,
  $cyc-CH_2N(phenyl)-$        13    19    62

$\underline{C_7H_4O}$, 104.0262  ext-ar(C=O),
  cyc-phenyl-CO-,  Y-phenyl-CO-,
  $cyc-C(-)(phenyl-O-)$        10    19    62

$\underline{C_6H_4N_2}$, 104.0373  $phenyl-arN_2$,  $ext-arN_2$        5    17    44

$\underline{C_6H_2NO}$, 104.0135  ON-phenyl-        3    14    44

| m/z, comp | Substructure, neighbor | Prop | Abnd | Spcf |
|---|---|---|---|---|

$\underline{C_4H_8O_3}$, 104.0473 $HOC_2H_4O-CO-CH_2-$,
$C_3H_7O-CO-O-$, $CH_3O-CO-CH(OCH_3)-$ — 2 16 73

$\underline{C_5H_2N_3}$, 104.0246 $ext-arN_3$ — 1 25 37
also $C_5H_{12}O_2$, 104.0836; $C_4H_8OS$, 104.0299
($-CH_2OCH_2CH_2SCH_2-$, $HOCH_2CH_2SCH_2CH_2-$)

| m/z 105 (38%) | | 9% | 37% | |
|---|---|---|---|---|

$\underline{C_7H_5O}$, 105.0340 — 18 47 67
  phenyl-CO-: -O- 25%, $CH_2$ 20%, -NH- 20%,
  CH 20% — 30 83 69
  cyc-CH(phenyl)-O-: -O- 75%, $CH_2$ 15%;
  $CH_2$ 35%, C=O 25%, CH 25% — 5 67 70
  also ext-ar(C=O), -phenyl-CO-, -phenyl-OCH$_2$-,
  phenyl-C(-)(OH)-

$\underline{C_8H_9}$, 105.0704 phenyl-CH($CH_3$)-,
  $CH_3$-phenyl-$CH_2$-, $(CH_3)_2$-phenyl etc — 25 34 80

$\underline{C_7H_7N}$, 105.0577 phenyl-N($CH_3$)-,
  -phenyl-$CH_2$-N(-)-, cyc-CH(phenyl)-NH-,
  phenyl-arN, $CH_3$-pyridyl-$CH_2$- — 7 29 56

$\underline{C_6H_3NO}$, 105.0214 pyridyl-CO-, -phenyl-NO,
  ext-arN(C=O) — 4 36+ 50

$\underline{C_6H_5N_2}$, 105.0451 phenyl-N=N-, $ext-arN_2$,
  arN amines — 3 23 45

$\underline{C_6HO_2}$, 104.9976 ar/unsatd/cyc
  C=O/-O-/-OH — 2 29 44

$\underline{C_3H_5S_2}$, 104.9839 (dithietane) — 1 55+ 87

| m/z, comp | Substructure, neighbor | Prop | Abnd | Spcf |
|---|---|---|---|---|

$\underline{C_5H_3N_3}$, 105.0325  ext-arN$_3$, pyridyl-N=N-
  etc                                             2    23   43

$\underline{C_4H_9O_3}$, 105.0551  C$_2$H$_5$O-CO-CH(OH)-,
  (CH$_3$O)$_3$C-, -CH$_2$-CH(OH)-CH(OH)-CH(OH)-,
  C$_3$H$_7$O-CO-O-, -CH$_2$OC(OC$_2$H$_5$)O-,
  C$_3$H$_7$OCH(-O-)O-                             1    30+  77

$\underline{C_5HN_2O}$, 105.0087  NC-pyrrolidonyl-,
  ar/ext-ar C=O/-OH/-NH-/-N=N- etc                1    26   34

$\underline{C_3H_5O_2S}$, 105.0013  cyc-CH$_2$CH(-CO-OH)S-,
  HO-CO-CH$_2$SCH$_2$-, -SCH$_2$CH$_2$-CO-O-       1    24   86
also C$_5$H$_{10}$Cl, 105.0470 (Cl(CH$_2$)$_5$-, ClC(C$_3$H$_7$)(CH$_3$)-);
  C$_4$H$_9$OS, 105.0377 (C$_3$H$_7$-CO-S-); C$_2$H$_2$Br, 104.9339
  (BrCH=CH-)

m/z 106 (27%)                                     7%   17%

$\underline{C_8H_{10}}$, 106.0782  CH$_3$-phenyl-CH$_2$- etc    21   13   76

$\underline{C_7H_6O}$, 106.0418  cyc-CH(phenyl)-O-,
  HO-phenyl-CH$_2$-, HO-(CH$_3$-)phenyl-,
  substd/cyc ketones                              15   13   62

$\underline{C_7H_8N}$, 106.0656  CH$_3$-pyridyl-CH$_2$-,
  (CH$_3$)$_2$-pyridyl-, cyc-CH$_2$N(phenyl)-,
  ar-amines                                       7    28   58

$\underline{C_6H_4NO}$, 106.0292  pyridyl-CO-,
  HO-(CH$_3$-)-pyridyl-, ar(-NH-CO-)              6    23   55

$\underline{C_6H_6N_2}$, 106.0529  ext-arN$_2$               5    13   47
also C$_5$H$_4$N$_3$, 106.0403 (arN$_3$); C$_5$H$_2$N$_2$O, 106.0165

| m/z, comp | Substructure, neighbor | Prop | Abnd | Spcf |
|---|---|---|---|---|

$(arN_2O, arN_2(C=O))$; $C_3H_6S_2$, 105.9917 $((CH_3S)_2CH-)$;
$C_4H_{10}O_3$, 106.0629 $(-CH_2CH(OH)CH(OH)CH(OH)-)$

| m/z 107 (31%) | | 15% | 24% | |
|---|---|---|---|---|

$\underline{C_7H_7}O$, 107.0496 $CH_3O$-phenyl-,
 HO-phenyl-$CH_2$-, cyc-CH(phenyl)-O-,
 substd/cyc/unsatd C=O/-OH ketones      16   27   65

$\underline{C_8H_{11}}$, 107.0860 polyunsatd/cyc hc      17   21   85

$\underline{C_6H_3O_2}$, 107.0133 ar/unsatd/cyc
 C=O/-O-/-OH       5   16   52

$\underline{C_6H_5}NO$, 107.0370 ON-phenyl-,
 $CH_3$-pyrrole-CO-       4   16   48

$\underline{C_7H_9}N$, 107.0734 $CH_3$-phenyl-NH-,
 phenyl-N($CH_3$)-, cyc/unsatd amines      4   16   55

$\underline{C_5H_5N_3}$, 107.0481 ext-$arN_3$      2   21   42
also $C_6H_7N_2$, 107.0607 (-NH-phenyl-NH-,
 $(CH_3)_2$-pyrazinyl-); $C_2H_4Br$, 106.9496 (BrCH($CH_3$)-
 etc); $C_5H_3N_2O$, 107.0244 (ext-$arN_2(C=O)$)

| m/z 108 (26%) | | 9% | 20% | |
|---|---|---|---|---|

$\underline{C_7H_8}O$, 108.0575 phenyl-$CH_2O$-,
 HO-phenyl-$CH_2$-, $CH_3$-phenyl-O-,
 $CH_3O$-phenyl-, phenyl-O-Y*-$CH_3$
 (Y* = -CO-O-), cyc/substd/unsatd
 C=O/-O-/HO-      18   24   61

$\underline{C_8H_{12}}$, 108.0938 polyunsatd/cyc hc      17   15   82

| m/z, comp | Substructure, neighbor | Prop | Abnd | Spcf |
|---|---|---|---|---|

$\underline{C_6H_4O_2}$, 108.0211 (HO-)$_2$phenyl-,
cyc/subtd/unsatd/ar C=O/-O-/HO-               7    18    46

$\underline{C_6H_6NO}$, 108.0448 $CH_3$O-pyridyl-,
$H_2$N-phenyl-O-, -NH-phenyl-O-,
substd/cyc -NH-CO-/-OH/-O- etc               4    20    50

$\underline{C_7H_{10}N}$, 108.0812 cyc/substd/unsatd
amines                                       4    17    57

$\underline{C_6H_8N_2}$, 108.0686 (CH$_3$)$_2$-pyrazinyl-,
phenyl-NHNH-, arN-NH- etc                    3    17    46
also $C_5H_4N_2O$, 108.0322 (-N-ar-NH-CO- etc); $C_5H_2NO_2$,
108.0084 (-O-ar-NH-CO-, $O_2$N-ar); $C_5H_6N_3$, 108.0559
(pyridyl-NHNH-, $H_2$N-(CH$_3$-)pyrimidinyl-); $C_4H_2N_3O$,
108.0195 ($H_2$N-pyrimidinyl-O-); $C_4H_4N_4$, 108.0433
(ext-arN$_4$): $C_6H_4S$, 108.0037 (arS); $C_6H_5P$, 108.0129
(ar-P)

m/z 109 (31%)                                14%   23%

$\underline{C_8H_{13}}$, 109.1017 polyunsatd/cyc hc        20    19    89

$\underline{C_7H_9O}$, 109.0653 1-decalones,
cyc/substd/unsatd ketones                    14    21    65

$\underline{C_6H_5O_2}$, 109.0289 HO-phenyl-O-,
$CH_3$-furyl-CO-                              6    19    49

$\underline{C_6H_7NO}$, 109.0526 -O-pyridyl-CH$_2$-,
$CH_3$O-pyridyl-, $H_2$N-phenyl-O-,
HO-phenyl-NH-                                3    27    48

$\underline{C_7H_{11}N}$, 109.0890 cyc/unsatd
imines/amines                                3    13    56

| m/z, comp | Substructure, neighbor | Prop | Abnd | Spcf |
|---|---|---|---|---|

$\underline{C_5H_3NO_2}$, 109.0163 ar-NO$_2$, arN-CO-OH,
-O-ar-NH-CO-

|  |  | 2 | 24 | 51 |
|---|---|---|---|---|

also $C_6H_5S$, 109.0152 (phenyl-S-); $C_5H_5N_2O$, 109.0400
(CH$_3$-pyrimidinyl-O-); $C_5HO_3$, 108.9925 (-O-ar-O-CO-);
$C_4H_3N_3O$, 109.0274 (H$_2$N-CO-imidazolyl); $C_3H_3Cl_2$,
108.9611; $C_6H_9N_2$, 109.0764 (arN-NH-); $C_5H_7N_3$,
109.0637 (H$_2$N-(CH$_3$)-pyrimidinyl)

| m/z 110 (25%) |  | 7% | 20% |  |
|---|---|---|---|---|

$\underline{C_7H_{10}O}$, 110.0731 substd/cyc/unsatd
ketones

|  |  | 17 | 20 | 60 |
|---|---|---|---|---|

$\underline{C_8H_{14}}$, 110.1094 substd/cyc/unsatd hc

|  |  | 20 | 13 | 82 |
|---|---|---|---|---|

$\underline{C_7H_{12}N}$, 110.0968 substd/cyc amines,
NC-C$_6$H$_{12}$-

|  |  | 7 | 26 | 63 |
|---|---|---|---|---|

$\underline{C_6H_6O_2}$, 110.0367 HO-phenyl-O-,
(HO)$_2$-phenyl-, substd/cyc
(C=O)$_2$/C=O/-O-

|  |  | 7 | 22 | 46 |
|---|---|---|---|---|

$\underline{C_6H_8NO}$, 110.0605 oxazoles,
cyc/substd/unsatd C=O/N/-CO-N-/C=N-OH,
ar-CO-NH-

|  |  | 4 | 24 | 41 |
|---|---|---|---|---|

$\underline{C_5H_4NO_2}$, 110.0241 pyrrolinedione-CH$_2$-,
maleimidyl-CH$_2$-, cyc C=O/-N-CO-/-O-

|  |  | 2 | 23 | 47 |
|---|---|---|---|---|

also $C_5H_2O_3$, 110.0003 (ar C=O/-O-/-CO-O-); $C_5H_6N_2O$,
110.0478 (arN$_2$-O-); $C_6H_{10}N_2$, 110.0842; $C_6H_6S$,
110.0193 (phenyl-S-); $C_4H_4N_3O$, 110.0352
(N$_2$ar(C=O)-NH$_2$); $C_4H_2N_2O_2$, 110.0114
(pyrazine-(-O-)$_2$- $C_6H_3Cl$, 109.9923 (Cl-phenyl-);
$C_2H_7O_3P$, 110.0132 ((CH$_3$O)$_2$-P(=O)-)

| m/z, comp | Substructure, neighbor | Prop | Abnd | Spcf |
|---|---|---|---|---|
| m/z 111 (28%) | | 10% | 20% | |

$\underline{C_8H_{15}}$, 111.1173 $(CH_3)_2$-cyclohexyl- etc    23    18    86

$\underline{C_7H_{11}O}$, 111.0809 cyclohexyl-CO-,
   $C_4H_9$CH=CH-CO-, substd/cyc/unsatd
   -CO-/-O-/-OH                    18    19    69

$\underline{C_6H_7O_2}$, 111.0445 -CO-$C_4H_8$-CO- (adipates),
   $CH_3$O-CO-CH=CHCH=CH-, cyc/unsatd
   C=O/(C=O)$_2$/-O-/-OH, ar-(OR)$_2$      8    23    54

$\underline{C_6H_9NO}$, 111.0683 cyc/unsatd
   amines/imines/C=O/-O-            3    21    47

$\underline{C_7H_{13}N}$, 111.1047 substd/cyc/unsatd
   amines, NC-$C_6H_{12}$-              3    16    62

$\underline{C_5H_3O_3}$, 111.0082 HO-pyrone-, furanoate,
   -CO-C(OH)=C($CH_3$)-CO-         3    14    57

$\underline{C_6H_7S}$, 111.0271 $(CH_3)_2$-thiophenyl-,
   thiophenyl-CH($CH_3$)-, ar-S-       2    31    73

$\underline{C_5H_5NO_2}$, 111.0319 pyrollyl-CO-O-,
   $H_2$N-CO-C($CH_3$)=CH-CO-, ar-O$CH_2$-CO-NH-    2    20    54
also $C_5H_7N_2O$, 111.0556 (cyc-carbamate, ar$N_2$-OH);
   $C_6H_{11}N_2$, 111.0920 (ar-N=CHN$(CH_3)_2$, cyc imines);
   $C_6H_4$Cl, 111.0001 (Cl-phenyl-); $C_4H_3N_2O_2$, 111.0193
   (-NH-CO-NH-CO-); $C_4H_5N_3O$, 111.0430 (HO-ar$N_2$-$NH_2$);
   $C_3H_5Cl_2$, 110.9767; $C_2HOCl_2$, 110.9403 ($Cl_2$CH-CO-,
   -$CCl_2$-CO-)

| m/z, comp | Substructure, neighbor | Prop | Abnd | Spcf |
|---|---|---|---|---|

| | | | | |
|---|---|---|---|---|
| m/z 112 (21%) | | 6% | 18% | |

$C_8\underline{H}_{16}$, 112.1251 $H-C_8H_{16}-Y*$, $H-C_8H_{16}-R-Y*$    19    9    82

$C_7\underline{H}_{12}O$, 112.0887 2-R-cyclohexanone-$CH_3$,
   $R-(CH_2)_6-CO-Y$, $C_3H_7CH=CH-CO-CH_2-$,
   cyc/subst/unsatd -CO-/-O-/-OH         12    18    61

$C_6\underline{H}_8O_2$, 112.0524 furyl-$CH(OCH_3)-$, cyc
   diketones, cyc/substd/unsatd
   -CO-O-/-CO-OH/-O-          10    20    54

$C_7\underline{H}_{14}N$, 112.1125 cyc/substd/unsatd
   amines (cyclohexyl-$NH-CH_2-$)      6    25    63

$C_6\underline{H}_{10}NO$, 112.0761
   cyc-$CH_2CH(N(C_2H_5)-CO-CH_3)-$, lactams,
   other cyc/unsatd amides, cyc HO-N=CH-,
   OCN-$C_5H_{10}-$           5    18    56

$C_5\underline{H}_4O_3$, 112.0160 furyl-CO-O-,
   ar(C=O)-CO-OH, HO-CO-C($CH_3$)=CH-CO-,
   ketoesters           3    26    53
also $C_5H_6NO_2$, 112.0397 (ketoamides); $C_5H_6NS$, 112.0223
   (arNS); $C_6H_{12}N_2$, 112.0998 (-N=N-, aminoimines);
   $C_5H_8N_2O$, 112.0635 (aminoamides); $C_6H_8S$, 112.0350
   ($CH_3$-thiophenyl-$CH_2-$); $C_4H_4N_2O_2$, 112.0271
   (-CH(-)-CO-NH-CO-NH-CH(-)-)

| | | | | |
|---|---|---|---|---|
| m/z 113 (23%) | | 6% | 21% | |

$C_8\underline{H}_{17}$, 113.1329 satd hc           19    12    87

$C_7\underline{H}_{13}O$, 113.0966 $C_6H_{13}-CO-$,

| m/z, comp | Substructure, neighbor | Prop | Abnd | Spcf |
|---|---|---|---|---|
| | cyc-CH(OH)CH$_2$CH(C$_3$H$_7$)-, -(CH$_2$)$_6$-CO-, | | | |
| | CH$_3$O-cyclohexyl- | 14 | 30+ | 71 |
| $\underline{C_6H_9O_2}$, 113.0602 | CH$_3$O-CO-CH=CH-CH$_2$CH$_2$-, | | | |
| | C$_3$H$_7$-CO-CH$_2$-CO-, -CH(OH)-(CH$_2$)$_4$-CO- | 11 | 16 | 60 |
| $\underline{C_5H_5O_3}$, 113.0238 | furyl-CO-O-, | | | |
| | CH$_3$O-CO-CH=CH-CO-, cyc C=O/-OH/-O- | 5 | 19 | 57 |
| $\underline{C_6H_{11}NO}$, 113.0839 | cycC=N-OH, | | | |
| | pyrrolidinyl-CO-CH$_2$-, cycN -OH/-O- | 3 | 25+ | 57 |
| $\underline{C_6H_9S}$, 113.0428 | thiabicycloalkane | 2 | 29 | 66 |
| $\underline{C_5H_7NO_2}$, 113.0475 | NC-CH(CH$_3$)-CO-O-CH$_2$-, | | | |
| | CH$_3$-succinimidyl-, glutarimidyl- | 2 | 22 | 52 |

also C$_7$H$_{15}$N, 113.1203 (cycN); C$_8$H$_3$N, 113.0264
(ext-arN); C$_6$H$_{13}$N$_2$, 113.1077 (cycN$_2$); C$_5$H$_9$N$_2$O,
113.0713 (pentanolactam-NH-); C$_5$H$_7$NS, 113.0301
(CH$_3$-thiazolyl-CH$_2$-); C$_4$H$_5$N$_2$O$_2$, 113.0349
(cyc(-N-CO-N-CO-)); C$_4$HO$_4$, 112.9874
(cyc-CH(-CO-O-)-CH(-CO-O-)-); C$_5$H$_5$OS, 113.0064
(CH$_3$O-thiophenyl-, CH$_3$S-furyl-, ar-SO-)

| m/z 114 (17%) | | | 11% | 10% |
|---|---|---|---|---|
| $\underline{C_6H_{10}O_2}$, 114.0680 | | | | |
| | cyc-CH(-CH$_2$-)(CH$_2$)$_3$-CO-O-, | | | |
| | CH$_3$O-CO-CH=C(CH$_3$)CH$_2$-, | | | |
| | CH$_3$O-CO-C$_4$H$_8$-Y* | 14 | 12 | 61 |
| $\underline{C_7H_{16}N}$, 114.1281 | (C$_3$H$_7$)$_2$NCH$_2$- etc | 11 | 15 | 64 |
| $\underline{C_5H_6O_3}$, 114.0316 | HO-CO-(CH$_2$)$_3$-CO- | 7 | 11 | 48 |

m/z, comp      Substructure, neighbor      Prop Abnd Spcf

$\underline{C_6H_{12}NO}$, 114.0917 $CH_3$-CO-NH-CH($C_3H_7$)-,
  $C_4H_9$-COCH(-)NH-, $C_4H_9$N($CH_3$)-CO-      5   17   56
also $C_7H_{14}O$, 114.1044 ($C_5H_{11}$-CO-$CH_2$-); $C_5H_8NO_2$,
  114.0554 ($CH_3$-CO-$CH_2$-N(-CO-$CH_3$)-); $C_6H_{14}N_2$, 114.1155
  ($C_2H_5$-imidazolidinyl-$CH_2$-); $C_4H_6N_2O_2$, 114.0427
  (-CH(-)-CO-NH-CO-NH-CH(-)-)

m/z 115 (34%)                              7%   27%

$\underline{C_9H_7}$, 115.0547                           28   24   87

$\underline{C_6H_{11}O_2}$, 115.0758 $CH_3$O-CO-($CH_2$)$_4$-,
  $C_3H_7$CH=CH-CO-O-, $CH_3$O-CO-$CH_2$C($CH_3$)$_2$-,
  ($CH_3$)$_2$-1,3-dioxanyl-             7   28   62

$\underline{C_5H_7O_3}$, 115.0394 $CH_3$O-CO-$CH_2$$CH_2$-CO-,
  $C_2H_5$O-CO-$CH_2$-CO-,
  $CH_3$-CO-CH(-)-CO-$OCH_2$-,
  -CO-$CH_2$CH($CH_3$)-O-CO-        5   38   55

$\underline{C_7H_{17}N}$, 115.1359 $C_3H_7$NHCH($C_3H_7$)-; and
  $\underline{C_8H_5N}$, 115.0421 ext-arN (indolyl-),
  pyrrolyl-ar                  7   22   60

$\underline{C_7H_{15}O}$, 115.1122 $C_6H_{13}$CH(OH)-,
  $C_5H_{11}$OCH($CH_3$)-, ($CH_3$)$_3$C-O-Y*-C($CH_3$)$_2$-;
  and $\underline{C_8H_3O}$, 115.0184 ext-ar(-CO-$CH_2$-),
  ext-ar-CO- etc             7   20   63
also $C_7H_3N_2$, 115.0295 (indazolyl-); $C_6H_{11}S$, 115.0584
  (cyc-$CH_2$$CH_2$CH(S$C_3H_7$)-) $C_5H_9NO_2$, 115.0632
  (cyc-N(-$CH_2$CH(-)$CH_3$)-CO-$OCH_2$-); $C_6H_{13}NO$, 115.0996
  (($C_2H_5$)$_2$N-CO-$CH_2$-); $C_4H_5NO_3$, 115.0268
  (cyc-C(OH)($CH_3$)-CO-NH-CO-); $C_4H_3O_4$, 115.0031

| m/z, comp | Substructure, neighbor | Prop | Abnd | Spcf |
|---|---|---|---|---|

$(HO-CO-CH_2-C(-)(-CO-OH)-)$; $C_5H_9NS$, 115.0458

$(SCN-(CH_2)_4-)$

---

m/z 116 (25%)      7%   16%

---

$\underline{C_9H_8}$, 116.0626 phenyl-$C_3H_3(-)-$, ext-ar    24   10   87

$\underline{C_8H_6N}$, 116.0499 indolyl, ext-arN,
NC-phenyl-$CH_2-$, phenyl-arN    11   15   60

$\underline{C_5H_8O_3}$, 116.0473 $CH_3O-CO-CH_2-CO-CH_2-$,
$CH_3O-CO-CH(-CO-CH_3)-$,
$CH_3O-CO-CH_2-CO-CH(-)-$    4   24   64

$\underline{C_8H_4O}$, 115.9898 ext-ar(C=O), ext-ar-CO-    7   11   64

$\underline{C_6H_{12}O_2}$, 116.0836 $CH_3O-CO-CH(C_3H_7)-$,
$-C(C_2H_5)_2-CO-O-$, $C_3H_7O-CO-CH(CH_3)-$,
$(C_2H_5O)_2C(CH_3)-$    6   13   62

$\underline{C_7H_4N_2}$, 116.0373 ext-arN$_2$
(benzimidazoles), phenyl-arN$_2$    3   12   49

also $C_5H_{10}NO_2$, 116.0710 ($CH_3O-CO-CH_2CH_2CH(NH_2)-$);
$C_4H_4O_4$, 116.0109 $((HO)_2-ar(-O-CO-)$,
$CH_3O-CO-C(-)_2-CO-O-$); $C_4H_6NO_3$, 116.0346
$(-C(CH_3)(OH)-CO-NH-CO-)$; $C_4H_4S_2$, 115.9761
(thiophenyl-S-, arS$_2$); $C_4H_8N_2O_2$, 116.0584
$(-CH_2NH-CO-CH_2NH-CO-)$; $C_4H_8O_2Si$, 116.0193
$((CH_3)_3SiO-CO-)$

---

m/z 117 (30%)      11%   20%

---

$\underline{C_9H_9}$, 117.0704 $CH_2=CH-phenyl-CH_2-$,
phenyl-$CH=CH-CH_2-$, indanyl- etc    24   17   83

| m/z, comp | Substructure, neighbor | Prop | Abnd | Spcf |
|---|---|---|---|---|

$\underline{C_8H_7N}$, 117.0577 $CH_3$-phenyl-CH=N-,
  ext-arN, cyc-$CH_2$CH(-)-N(phenyl)-,
  phenyl-arN

| | | 9 | 16 | 61 |
|---|---|---|---|---|

$\underline{C_8H_5O}$, 117.0340 ext-ar(C=O), HO-ext-ar,
  ext-ar-CO-

| | | 6 | 16 | 63 |
|---|---|---|---|---|

$\underline{C_5H_9O_3}$, 117.0551 -$CH_2$O($CH_2$)$_3$-CO-O-,
  $CH_3$O-CO-$CH_2$-CO-$CH_2$-,
  -CH(-CHO-)-OC($CH_3$)$_2$-O-,
  $CH_3$O-CO-$CH_2$CH(O$CH_3$)-

| | | 4 | 18 | 67 |
|---|---|---|---|---|

$\underline{C_7H_5N_2}$, 117.0451 ext-arN$_2$, phenyl-arN$_2$

| | | 2 | 15 | 46 |
|---|---|---|---|---|

$\underline{C_6H_{13}O_2}$, 117.0915 $C_5H_{11}$-CO-O-,
  $CH_3$OCH($CH_3$)CH($CH_3$)CH(OH)-,
  -$CH_2$O-($CH_2$)$_4$-OCH$_2$-, $C_3H_7$OCH(O$C_2H_5$)-

| | | 2 | 12 | 61 |
|---|---|---|---|---|

also $C_6H_{13}$S, 117.0741 ($C_4H_9$SCH($CH_3$)-); $C_5H_{13}$OSi,
117.0635 (($CH_3$)$_3$SiOCH($CH_3$)-, $C_3H_7$OSi($CH_3$)$_2$-);
$CCl_3$, 116.9064; $C_7H_3$NO, 117.0214 (ext-arN(C=O),
ext-arN-CO-); $C_5H_9$OS, 117.0377 ($CH_3$-1,3-oxathiane);
$C_2HF_3Cl$, 116.9718 (CF$_3$C(-)Cl-)

| m/z 118 (23%) | | 6% | 18% | |
|---|---|---|---|---|

$\underline{C_9H_{10}}$, 118.0782 phenyl-$C_3H_5$(-)-,
  tetralins, -$CH_2$phenyl-$C_2H_4$-

| | | 22 | 15 | 81 |
|---|---|---|---|---|

$\underline{C_8H_8N}$, 118,0656 phenyl-cyc-amine
  (-CH($CH_3$)-N(phenyl)-),
  phenyl-C(=NCH$_3$)-, phenyl-CH=NCH$_2$-

| | | 10 | 17 | 61 |
|---|---|---|---|---|

$\underline{C_8H_6O}$, 118.0418 phenyl-$CH_2$-CO-,
  -CH(phenyl)-CH(-)-O-, ext-ar(-CO-$CH_2$-)

| | | 9 | 19 | 59 |
|---|---|---|---|---|

| m/z, comp | Substructure, neighbor | Prop | Abnd | Spcf |
|---|---|---|---|---|

$\underline{C_7H_6N_2}$, 118.0529 ext-ar-$N_2$, phenyl-ar$N_2$,
phenyl-C(-N-)=N-                                    6    17    54

$\underline{C_7H_4NO}$, 118.0292 cyc-N(phenyl)-CO-,
-phenyl-CH=N-O-, HO-ext-arN,
$\underline{o}$-$H_2$N-CO-phenyl-                            4    18    56
also $C_7H_2O_2$, 118.0054 (ext-ar(C=O)-O-, ar-CO-O-,
ar-(C=O)$_2$-); $C_6H_4N_3$, 118.0403 (ext-ar$N_3$); $C_5H_{10}O_3$,
118.0629 (HO-CO-C(OH)($C_3H_7$)-, $C_4H_9$O-CO-O-,
HOC$_3H_6$O-CO-CH$_2$-); $C_6H_2N_2$O, 118.0165 (NC-pyridyl-O-);
$C_4H_6O_4$, 118.0265 (HO-CO-C(-)(CH$_3$)-); $C_4H_8NO_3$,
118.0503 (cyc-CH(OH)-CH(OH)-N(CH$_3$)-CO-,
$H_2$N-CO-OCH$_2$CH(OH)CH$_2$-); $C_5H_{12}NO_2$, 118.0867
($C_4H_9$NH-CO-O-, $C_4H_9$O-CO-NH-)

| m/z 119 (30%) | | 12% | 21% | |
|---|---|---|---|---|

$\underline{C_9H_{11}}$, 119.0860 phenyl-C(CH$_3$)$_2$-,
(CH$_3$)$_3$-phenyl- etc                              22    17    79

$\underline{C_8H_7O}$, 119.0496 CH$_3$-phenyl-CO-,
dihydrobenzofurans, -CH$_2$CH(phenyl)-O-,
substd/cyc/unsatd ketones                          10    20    60

$\underline{C_7H_5NO}$, 119.0370 $H_2$N-phenyl-CO-,
phenyl-NH-CO-, cyc-N(phenyl)-CO-,
ext-ar(-NH-CO-)                                      6    21    59

$\underline{C_7H_3O_2}$, 119.0133 ar(C=O)$_2$, -phenyl-CO-O-,
phenyl-ar(-O-CO-)                                    4    21    54

$\underline{C_8H_9N}$, 119.0734 benzo-NHCH$_2$-,
phenyl-CH=N-CH$_2$-, phenyl-CH$_2$CH(NH$_2$)-,
cyc-CH$_2$CH(phenyl)N(-)-                             5    13    54

MASS SPECTRAL CORRELATIONS

| m/z, comp | Substructure, neighbor | Prop | Abnd | Spcf |
|---|---|---|---|---|

$C_7H_7N_2$, 119.0607 $CH_3$-phenyl-N=N-,
phenyl-CH=N-NH-, $CH_3$CH=CH-pyrazyl- — 3  19  43
also $C_6H_5N_3$, 119.0481 (arN$_3$); $C_6H_3N_2O$, 119.0244
  (imidazolidone-ar, HO-ext-arN$_2$); $C_5H_3N_4$, 119.0355
  (ext-arN$_4$); $C_2F_5$, 118.9920

| m/z 120 (24%) | | 7% | 19% | |
|---|---|---|---|---|

$C_9H_{12}$, 120.0938 $(CH_3)_2$-phenyl-$CH_2$- etc   17  14  73

$C_8H_8O$, 120.0575 phenyl-CO-$CH_2$-,
  cyc-$CH_2$CH(phenyl)-O-, substd/cyc
  ketones   12  18  58

$C_8H_{10}N$, 120.0812 $(CH_3)_3$-pyridyl-,
  $H_2$NCH($CH_2$-phenyl)-,
  ar/unsatd/cyc/substd amines   6  30  64

$C_7H_6NO$, 120.0448 $H_2$N-phenyl-CO-,
  phenyl-NH-CO-, HO($CH_3$-)pyridyl-$CH_2$-   7  25  56

$C_7H_4O_2$, 120.0211 HO-phenyl-CO-,
  HO-CO-phenyl-, -phenyl-CO-O-,
  -phenyl-O-CO-, ext-ar(-O-CO-)
  (dihydrocoumarins)   6  23  53

$C_7H_8N_2$, 120.0686 $C_3H_7$-pyrazines,
  phenyl-CH=N-NH-, ar-amines   3  23  43

$C_6H_4N_2O$, 120.0322 ar(-NH-CO-NH-),
  ext-arN$_2$O   3  18  55
also $C_5H_4N_4$, 120.0433 (ext-arN$_4$); $C_6H_6N_3$, 120.0559
  (ext-arN$_3$); $C_6H_2NO_2$, 120.0084 (ext-ar(-CO-NH-CO-),
  -pyridyl-CO-O-, -CO-NH-ar-CO-); $C_5H_2N_3O$, 120.0195

| m/z, comp | Substructure, neighbor | Prop | Abnd | Spcf |
|---|---|---|---|---|

(ext-arN$_3$(C=O)); C$_3$H$_5$Br, 119.9574
(cyc-CH$_2$CH(-)CHBr-); C$_2$HOBr, 119.9210 (=CBr-CO-,
arBr(OH))

| m/z 121 (30%) | | 11% | 24% | |
|---|---|---|---|---|

C$_9$H$_{13}$, 121.1017 polyunsatd/cyc hc — 20  27  85

C$_8$H$_9$O, 121.0653 cyc-CH(-)CH(phenyl)O-,
phenyl-C(-)(OCH$_3$)-, phenyl-CHY*-Y-OCH$_3$,
substd/cyc/unsatd C=O/-O-/-OH — 13  27  62

C$_7$H$_5$O$_2$, 121.0289 HO-phenyl-CO-,
CH$_3$(-CH$_2$-)furyl-CO-, furyl-CH=CH-CO-,
-O-phenyl-CH$_2$O-, substd/cyc C=O/-O- — 11  30  58

C$_7$H$_7$NO, 121.0526 phenyl-CH=N-O-,
phenyl-NH-CO-, cyc/unsatd/substd
amines/C=O/-O-/-OH — 4  14  52

C$_8$H$_{11}$N, 121.0890 (CH$_3$)$_2$phenyl-NH-,
ar-amines — 3  13  55

also C$_6$H$_3$NO$_2$, 121.0163 (O$_2$N-phenyl-, ar/arN
C=O/-NH/-O-/OH); C$_6$H$_5$N$_2$O, 121.0400 (pyrazones,
HO-phenyl-N=N-); C$_7$H$_9$N$_2$, 121.0764 ((CH$_3$)$_2$N-pyridyl-,
pyrazines); C$_5$H$_3$N$_3$O, 121.0274 (HO-ext-arN$_3$-);
C$_7$H$_5$S, 121.0115 (cyc-CH(phenyl)-S-, arS); C$_6$H$_7$N$_3$,
121.0637 (ext-arN$_3$); C$_2$H$_2$BrO, 120.9289 (BrCH$_2$-CO-)

| m/z 122 (23%) | | 7% | 17% | |
|---|---|---|---|---|

C$_8$H$_{10}$O, 122.0731 C$_2$H$_5$O-phenyl-,
phenyl-O-Y*-C$_2$H$_5$, cyc/substd/unsatd
ketones/-OH — 14  19  60

# MASS SPECTRAL CORRELATIONS

| m/z, comp | Substructure, neighbor | Prop | Abnd | Spcf |
|---|---|---|---|---|

$\underline{C}_9\underline{H}_{14}$, 122.1095 polyunsatd/cyc hc     16   13   81

$\underline{C}_7\underline{H}_6\underline{O}_2$, 122.0367 $CH_3O$-(HO-)phenyl-,
(HO-)$_2$phenyl-$CH_2$-, cyc/substd C=O/-O-    11   14   54

$\underline{C}_7\underline{H}_8\underline{N}O$, 122.0605 ext-ar(NH)(C=O),
$CH_3O$-phenyl-NH-, substd/cyc
amines/C=O/-O-       4   18   48

$\underline{C}_8\underline{H}_{12}\underline{N}$, 122.0968 cyc/substd/unsatd amines   4   18   59

$\underline{C}_6\underline{H}_4\underline{N}O_2$, 122.0241 pyridone-CO-,
HO-CO-pyridyl-, H-CO-pyrrolyl-$CH_2O$-,
$O_2$N-phenyl-       3   23   50
also $C_6H_2O_3$, 122.0003; $C_6H_6N_2O$, 122.0478
(HO-($CH_3$-)$_2$-pyrazinyl-); $C_7H_{10}N_2$, 122.0842
(($CH_3$)$_2$pyrazinyl-$CH_2$-); $C_5H_4N_3O$, 122.0352
(HO-ext-arN$_3$-); $C_5H_2N_2O_2$, 122.0114; $C_7H_6S$, 122.0193
(ext-arS)

| m/z 123 (25%) | | 9% | 22% | |
|---|---|---|---|---|

$\underline{C}_9\underline{H}_{15}$, 123.1173 polyunsatd/cyc hc    20   17   88

$\underline{C}_8\underline{H}_{11}O$, 123.0809 cyc/substd/unsatd
C=O/-OH        14   20   64

$\underline{C}_7\underline{H}_7O_2$, 123.0445 phenyl-CO-O-,
HO-phenyl-CO(OH)-, phenyl-C(-)(-O-)$_2$,
ar/unsatd/cyc C=O/-OH/-O-    9   21   54

$\underline{C}_6\underline{H}_3O_3$, 123.0082 ar/unsatd/cyc
-CO-O-/-CO-/-O-/-OH     3   17   55

$\underline{C}_6\underline{H}_5\underline{N}O_2$, 123.0319 $O_2$N-phenyl-,

| m/z, comp   Substructure, neighbor | Prop | Abnd | Spcf |
|---|---|---|---|
| HO-(CH₃)-pyridonyl- | 3 | 18 | 45 |

$\underline{C_7H_9NO}$, 123.0683 CH₃O-phenyl-NH-,
ar-CH₂CH(-CO-NH₂)-,

| CH₃O-(H₂N-)-phenyl- | 2 | 21 | 45 |

also C₈H₁₃N, 123.1047 (cyc/unsatd amines); C₇H₇S,
123.0271 (phenyl-SCH₂-, CH₃S-phenyl-); C₆H₇N₂O,
123.0556 (CH₃-pyrimidinyl-OCH₂-,
CH₃O-(CH₃-)-pyrazinyl-); C₅H₅N₃O, 123.0430

| m/z 124 (19%) | | 6% | 22% |
|---|---|---|---|

| $\underline{C_8H_{12}O}$, 124.0887 substd/cyc/unsatd ketones | 14 | 18 | 60 |

| $\underline{C_9H_{16}}$, 124.1251 substd/cyc/unsatd hc | 15 | 10 | 83 |

$\underline{C_7H_8O_2}$, 124.0524 CH₃O-phenyl-O-,
(HO)₂-phenyl-CH₂-, cyc/substd/unsatd

| C=O/-O- | 8 | 24 | 48 |

$\underline{C_8H_{14}N}$, 124.1125 substd/cyc amines,

| NC-C₇H₁₄- | 6 | 30 | 60 |

| $\underline{C_7H_8S}$, 124.0350 CH₃-phenyl-S- | 2 | 59 | 71 |

$\underline{C_7H_{10}NO}$, 124.0761 oxazoles, cyc/substd

| amines/-O-/C=N-OH | 3 | 24 | 40 |

| $\underline{C_6H_6NO_2}$, 124.0397 furyl-CO-CH(NH₂)-, maleimides, cyc/unsatd, amines/C=O/-O- | 3 | 17 | 48 |

| $\underline{C_6H_4O_3}$, 124.0160 unsatd/ar C=O/-O-/-OH | 3 | 13 | 44 |

also C₆H₈N₂O, 124.0635; C₆H₆NS, 124.0223
(pyridyl-SCH₂-, CH₃S-pyridyl-, arS-NH-); C₅H₄N₂O₂,
124.0271 (O₂N-pyridyl-, CH₃-pyrazinyl-(-O-)₂-

| m/z, comp | Substructure, neighbor | Prop | Abnd | Spcf |
|---|---|---|---|---|

| m/z 125 (22%) | | 6% | 20% | |
|---|---|---|---|---|

$\underline{C_9H_{17}}$, 125.1329 cyc/unsatd hc — 21, 19, 87

$\underline{C_8H_{13}O}$, 125.0966 cyclohexyl-$CH_2$-CO-,
-$C_7H_{14}$-CO-, cyc/unsatd/substd
-CO-/-O-/-OH; and $\underline{C_9HO}$, 125.0027
ext-ar(C=O) — 11, 16, 66

$\underline{C_7H_9O_2}$, 125.0602 -CO-$C_5H_{10}$-CO-,
cyc/unsatd C=O/-O-/-OH, ar-$(OCH_3)_2$ — 8, 19, 58

$\underline{C_6H_5O_3}$, 125.0238 $CH_3$O-pyrone-,
furanoates etc — 4, 15, 56

$\underline{C_7H_9S}$, 125.0428 $C_3H_7$-thiophenyl-, ar-SR — 2, 20, 72

$\underline{C_6H_7NO_2}$, 125.0475 ar-NH-CO-OR,
furyl-CO-$NHCH_2$-, $CH_3$-ar-$NO_2$ — 2, 20, 55
also $C_7H_{11}NO$, 125.0839 (arN-O-); $C_8H_{15}N$, 125.1203;
$C_6H_9N_2O$, 125.0713 (cyc carbamates); $C_7H_6Cl$, 125.0157
(Cl-phenyl-$CH_2$-); $C_6H_7NS$, 125.0301 (arN-S-);
$C_7H_{13}N_2$, 125.1077 (cyc imines); $C_5HO_4$, 124.9874
($-CH_2$O-CO-C≡C-CO-O-, $(-O)_2$ar-CO-); $C_6H_4NCl$,
125.0031 (Cl-ar-NH-)

| m/z 126 (20%) | | 6% | 20% | |
|---|---|---|---|---|

$\underline{C_9H_{18}}$, 126.1407 H-$C_9H_{18}$-Y*, H-$C_9H_{18}$-R-Y*;
and $\underline{C_{10}H_6}$, 126.0469 naphthyl — 20, 11, 85

$\underline{C_8H_{14}O}$, 126.1044 2-R-cyclopentanone-$C_3H_7$,
R-$(CH_2)_7$-CO-Y, cyc/substd/unsatd
-CO-/-O-/-OH; and $\underline{C_9H_2O}$, 126.0105

| m/z, comp | Substructure, neighbor | Prop | Abnd | Spcf |
|---|---|---|---|---|
| | ext-ar(C=O) | 11 | 21 | 62 |

$\underline{C_7H_{10}O_2}$, 126.0680 cyc/substd/unsatd
-CO-/-CO-O-/-O- etc

| | | 7 | 19 | 52 |
|---|---|---|---|---|

$\underline{C_8H_{16}N}$, 126.1281 cyc/substd/unsatd
amines; and $\underline{C_9H_4N}$, 126.0343 substd
quinolines

| | | 5 | 17 | 60 |
|---|---|---|---|---|

$\underline{C_7H_{12}NO}$, 126.0917 cyc-$C_5H_9$-N(-CO-CH$_3$)-,
cyc/unsatd amides, oximes, isocyanates

| | | 4 | 16 | 54 |
|---|---|---|---|---|

$\underline{C_6H_6O_3}$, 126.0316 cyc/substd/unsatd
C=O/-CO-O-/-OH
(cyc-$CH_2CH_2CH$(-CO-)CH(-CO-OH)-,
HO-pyrone-$CH_2$-)

| | | 3 | 23 | 50 |
|---|---|---|---|---|

also $C_6H_8NO_2$, 126.0554 (cyc-CO-NR-CO-, ar-NH-CO-OCH$_3$);
$C_7H_{10}S$, 126.0506 (arS); $C_6H_8NS$, 126.0380 (arNS);
$C_7H_{14}N_2$, 126.1155 (unsatd diamines); $C_5H_6N_2O_2$,
126.0427 (unsatd/cyc -NH-CO-N(-)-CO-) $C_6H_{10}N_2O$,
126.0791 (C=N-NH-CO-)

| m/z 127 (26%) | | 7% | 16% |
|---|---|---|---|

$\underline{C_{10}H_7}$, 127.0547 naphthyl-Y*, ext-ar; and
$\underline{C_9H_{19}}$, 127.1486 satd hc

| | | 25 | 11 | 87 |
|---|---|---|---|---|

$\underline{C_7H_{11}O_2}$, 127.0758 $C_4H_9$-CO-CH$_2$-CO-,
CH$_3$O-(Y*-)cyclohexyl-O-,
CH$_3$O-CO-CH=C(CH$_3$)CH$_2$CH$_2$-

| | | 8 | 14 | 62 |
|---|---|---|---|---|

$\underline{C_8H_{15}O}$, 127.1122 $C_7H_{15}$-CO-,
-CH(OH)-(CH$_2$)$_7$-; and $\underline{C_9H_3O}$, 127.0184
ext-ar(C=O)(-benzofuranonyl-),
-O-phenyl-ar-O-

| | | 9 | 14 | 66 |
|---|---|---|---|---|

| m/z, comp | Substructure, neighbor | Prop | Abnd | Spcf |
|---|---|---|---|---|

$\underline{C_6H_7O_3}$, 127.0394 ar/cyc/unsatd
  -O-/-OH/C=O

|  |  | 4 | 18 | 58 |
|---|---|---|---|---|

I, 126.9043 iodo

|  |  | 3 | 21 | 68 |
|---|---|---|---|---|

$\underline{C_5H_7N_2O_2}$, 127.0505
  $-CH_2-CO-NHCH(CH_3)-CO-NH-$,
  $-C(-)_2-CO-NH-CH_2-CO-NH-CH_2-$

|  |  | 2 | 30 | 61 |
|---|---|---|---|---|

$\underline{C_6H_9NO_2}$, 127.0632 arN -O-/-OH,
  $-N(C_2H_5)-CO-C(CH_3)=C(-)-O-$

|  |  | 2 | 24 | 62 |
|---|---|---|---|---|

also $C_9H_5N$, 127.0421 (ext-arN); $C_7H_{15}N_2$, 127.0295
  (ext-$arN_2$); $C_7H_{13}NO$, 127.0196 (piperidyl-$CO-CH_2-$);
  $C_7H_{11}S$, 127.0584 (thiabicycloalkane); $C_6H_9NS$,
  127.0457 ($C_2H_5$-thiazolyl-$CH_2-$); $C_5H_3O_4$, 127.0031;
  $C_6H_6NCl$, 127.0188 (Cl-phenyl-NH-)

m/z 128 (25%)

|  |  |  | 8% | 17% |
|---|---|---|---|---|

$\underline{C_{10}H_8}$, 128.0626 ext-ar hc

|  |  | 22 | 13 | 85 |
|---|---|---|---|---|

$\underline{C_7H_{12}O_2}$, 128.0836 $C_2H_5O-CO-CH=C(CH_3)CH_2-$,
  $CH_3O-(HO-)$-cyclohexyl-,
  $C_2H_5O-CO-(CH_2)_4-$,
  $CH_3O-CO-C(CH_3)_2CH_2CH_2-$

|  |  | 6 | 17 | 54 |
|---|---|---|---|---|

$\underline{C_9H_6N}$, 128.0499 quinolinyl, phenyl-arN

|  |  | 8 | 11 | 61 |
|---|---|---|---|---|

$\underline{C_9H_4O}$, 128.0262 ext-ar(C=O)
  (benzofuranonyl-), ext-ar-CO-,
  phenyl-cyc(C=O)

|  |  | 7 | 12 | 58 |
|---|---|---|---|---|

$\underline{C_7H_{14}NO}$, 128.1074 $(C_3H_7)_2N-CO-$,
  $CH_3-CO-NH-CH(C_3H_7)-$, $C_6H_{13}NH-CO-$,

| m/z, comp | Substructure, neighbor | Prop | Abnd | Spcf |
|---|---|---|---|---|

$\underline{cyc-CH_2CH_2C(CH_3)(OH)CH_2CH_2N(CH_3)}$-     3   28   56

$\underline{C_6H_8O_3},\ 128.0473$ $CH_3O-CO-C(OCH_3)=CHCH_2-$,
 $CH_3O-CO-(CH_2)_3-CO-$,
 $C_2H_5O-CO-CH(-CO-CH_3)-$     3   20   50
also $C_6H_{10}NO_2$, 128.0710 (($CH_3)_2NC(OCH_3)=CH-CO-$,
 $cyc-CH_2CH_2CH(-NH-CO-OC_2H_5)$); $C_8H_{16}O$, 128.1200
 ($C_6H_{13}-CO-CH_2-$); $C_7H_{16}N_2$, 128.1311 (($C_3H_7)_2C=N-NH-$);
 $C_8H_4N_2$, 128.0373 (ext-arN$_2$(-quinoxalinyl));
 $C_5H_6NO_3$, 128.0346 ($-CH_2CH(NH-CO-CH_3)-CO-O-$);
 $C_6H_5OCl$, 128.0029 (Cl-phenyl-O-); $C_5H_4O_4$, 128.0109
 ($-CH_2O-CO-)_2CH-$)

m/z 129 (28%)     10%   20%

$\underline{C_{10}H_9},\ 129.0704$ indenes, indanes,
 tetralins etc     21   16   84

$\underline{C_7H_{13}O_2},\ 129.0915$ $HO-CO-cyclohexyl-$,
 $HO-CO-(CH_2)_6-$, $CH_3-CO-(CH_2)_5-$,
 $-(CH_2)_6-CO-O-$     7   20   61

$\underline{C_8H_{17}O},\ 129.1278$ $C_7H_{15}CH(OH)-$,
 $C_6H_{13}OCH(CH_3)-$; and $\underline{C_9H_5O},\ 129.0340$
 phenyl-CH=C(-)-CO-, -indonyl-O-,
 $cyc-CH_2C(-)(phenyl)-$     8   15   64

$\underline{C_9H_7N},\ 129.0577$ indolyl-CH$_2$-,
 1,2-dihydroquinolinyl-, 2-quinolinyl-     8   14   61

$\underline{C_6H_9O_3},\ 129.0551$ $-CO-(CH_2)_4-CO-O-$,
 $CH_3O-CO-(CH_3)_2-CO-$,
 $-C(OCH_3)=CHCH_2CH_2-CO-O-$,
 $-CH_2CH(OH)CH_2O-CO-CH_2CH_2-$     4   19   59

MASS SPECTRAL CORRELATIONS

m/z, comp    Substructure, neighbor    Prop Abnd Spcf

$\underline{C_8H_5N_2}$, 129.0451 ext-arN$_2$ (quinazolinyl,
  CH$_3$-benzimidazolyl-)                4   18   53
also C$_8$H$_3$NO, 129.0214 (ext-arN-CO-, ext-arN(C=O));
  C$_5$H$_5$O$_4$, 129.0187 (-O-CO-CH$_2$CH(OH)CH$_2$-CO-)
  C$_6$H$_{13}$N$_2$O, 129.1026 (-NH(CH$_2$)$_3$N(-CO-CH$_3$)-CH$_2$-);
  C$_7$H$_3$N$_3$, 129.0325 (ext-arN$_2$-N(-)-); C$_7$H$_{13}$S, 129.0741
  (cyc-CH$_2$CH$_2$CH(-SC$_4$H$_9$)-)

m/z 130 (22%)                          6%  19%

$\underline{C_{10}H_{10}}$, 130.0782 phenyl-C$_4$H$_5$(-)-,
  C$_5$H$_9$-phenyl-CH$_2$-, ext-ar       18   15   83

$\underline{C_9H_8N}$, 130.0656 ext-arN (CH$_3$-indolinyl-,
  indole-CH$_2$-), phenyl-arN        9   26   61

$\underline{C_9H_6O}$, 130.0418 ext-ar(C=O),
  -phenyl-CH$_2$CH(-)-CO-          8   18   65

$\underline{C_8H_6N_2}$, 130.0529 ext-arN$_2$
  (quinazolinyl), phenyl-arN$_2$       5   18   48

$\underline{C_8H_4NO}$, 130.0292 NC-phenyl-CO-,
  ext-arN-CO-, ext-arN-O-        3   23   48
also C$_7$H$_{14}$O$_2$, 130.0993 (CH$_3$O-CO-C(CH$_3$)(C$_3$H$_7$)-);
  C$_6$H$_{10}$O$_3$, 130.0629 (C$_2$H$_5$O-CO-CH(-CO-CH$_3$)-);
  C$_6$H$_{12}$NO$_2$, 130.0866 (CH$_3$O-CO-(CH$_2$)$_3$-CH(NH$_2$)-);
  C$_5$H$_8$NO$_3$, 130.0503 (-N(-CO-O-C$_2$H$_5$)-CH$_2$CH$_2$O-,
  -O-CO-CH$_2$CH$_2$CH(NH$_2$)-CO-); C$_6$H$_{14}$N$_2$O, 130.1104
  (H$_2$NCH$_2$CH$_2$N(C$_2$H$_4$OH)-CH$_2$CH$_2$-); C$_5$H$_{10}$N$_2$O$_2$, 130.0740
  ((-CH$_2$-)$_2$C=N-NH-CO-OCH$_3$); C$_3$H$_2$N$_2$O$_4$, 130.0013
  ((O$_2$N)$_2$-ar)

| m/z, comp | Substructure, neighbor | Prop | Abnd | Spcf |
|---|---|---|---|---|

m/z 131 (27%)                                                  10%   20%

$\underline{C}_{10}\underline{H}_{11}$, 131.0860 $C_3H_5$-phenyl-$CH_2$-,
  tetralinyl etc                                               18    20    82

$\underline{C}_9\underline{H}_7O$, 131.0496 phenyl-CH=CH-CO-,
  ext-ar(C=O), phenyl-CH(-)$CH_2$-CO-                          11    23    68

$\underline{C}_9\underline{H}_9N$, 131.0734 ext-arN, phenyl-arN   5    14    60

$\underline{C}_8\underline{H}_7\underline{N}_2$, 131.0607 $CH_3$-indazolyl,
  phthalazinyl                                                  3    22    57

$\underline{C}_7\underline{H}_{15}\underline{O}_2$, 131.1071 $(C_3H_7O)_2$CH-,
  $C_3H_7$-CH(-O-CO-$CH_3$)-$CH_2$-; and
  $\underline{C}_8\underline{H}_3\underline{O}_2$, 131.0133 -O-ar-CO-CH=CH-,
  ext-ar(C=O)-O-                                                3    15    56

$\underline{C}_8\underline{H}_5NO$, 131.0370 ext-arN(C=O),
  phenyl-arNO etc                                               3    14    47
also $C_6H_{11}O_3$, 131.0707 (HOCH$_2$-(CH$_3$-)$_2$-dioxetanyl);
  $C_7H_3N_2O$, 131.0244 (ext-arN$_2$-O-); $C_6H_{11}OS$, 131.0533
  ((CH$_3$)$_2$-1,3-oxathiane); $C_5H_9NO_3$, 131.0581
  (CH$_3$-CO-NH-CH(-CO-O-CH$_3$)-); $C_6H_{15}OSi$, 131.0791
  ((CH$_3$)$_3$Si-O-CH($C_2H_5$)-); $C_3F_5$, 130.9920

m/z 132 (21%)                                                   6%    17%

$\underline{C}_{10}\underline{H}_{12}$, 132.0938 tetralins,
  phenyl-$C_4H_7$(-)- etc                                      16    15    80

$\underline{C}_9\underline{H}_8O$, 132.0575 ext-ar(C=O),
  -CH(CH$_3$)-C(-)(phenyl)-O-,
  (CH$_3$)$_2$-phenyl-CO-                                      11    17    64

| m/z, comp | Substructure, neighbor | Prop | Abnd | Spcf |
|---|---|---|---|---|
| $C_9H_{10}N$, 132.0812 cyc-$CH_2CH_2C(NH_2)$(phenyl)-, ar amines/imines, ext-arN | | 7 | 24 | 63 |
| $C_8H_6NO$, 132.0448 $CH_3$-phenyl-NH-CO-, HO-ext-arN | | 6 | 15 | 60 |
| $C_8H_8N_2$, 132.0686 phenyl-arN$_2$ | | 4 | 16 | 50 |
| $C_8H_4O_2$, 132.0211 ext-ar(C=O)$_2$, HO-CO-ext-ar, ext-ar(-CO-O-) | | 4 | 16 | 53 |
| $C_5H_8O_4$, 132.0422 ($CH_3$O-CO-)$_2$CH-, (HO-CO-)$_2$C($C_2H_5$)- | | 2 | 27 | 63 |
| $C_7H_4N_2O$, 132.0322 ext-arN$_2$-O- (benzoimidazolyl-O-), ext-arN$_2$(C=O) | | 2 | 18 | 49 |

also $C_7H_6N_3$, 132.0559 (ext-arN$_3$); $C_2F_2Cl_2$, 131.9344

| m/z 133 (27%) | | 11% | 19% | |
|---|---|---|---|---|
| $C_{10}H_{13}$, 133.1017 $CH_3$-phenyl-C($CH_3$)$_2$-, ($CH_3$)$_4$-phenyl- etc | | 17 | 19 | 76 |
| $C_9H_9O$, 133.0653 ($CH_3$)$_2$-phenyl-CO-, dyhydrobenzopyrans, phenyl-$CH_2CH_2$-CO- | | 9 | 23 | 61 |
| $C_8H_5O_2$, 133.0289 -phenyl-CO-OCH$_2$-, -$CH_2$-phenyl-O-CO-, -CH(OH)-C(-)(phenyl-O-)- | | 5 | 23 | 56 |
| $C_8H_7NO$, 133.0526 $CH_3$-phenyl-NH-CO-, phenyl-arON | | 4 | 18 | 55 |
| $C_9H_{11}N$, 133.0890 benzo-cycN | | 4 | 11 | 53 |

m/z, comp     Substructure, neighbor          Prop Abnd Spcf

$\underline{C_8H_9N_2}$, 133.0764 $(CH_3)_2$-pyrazinyl-CH=CH-,
$(CH_3)_2$N-phenyl-N(-)-                              2    23    52
also $C_7H_5N_2O$, 133.0400 (phenyl-ON=N-$CH_2$-,
-phenyl-NHNH-CO-); $C_7H_3NO_2$, 133.0163 (ext-arN(C=O)$_2$);
$C_6H_5N_4$, 133.0511 (arN$_4$); $C_4H_6Br$, 132.9652
(BrCH$_2$CH=CHCH$_2$-); $C_8H_6P$, 133.0207 (benzo-cycP);
$C_4H_5O_5$, 133.0136 (-CH(OH)-CH(OH)-CH(OH)-CH(OH)-O-)

m/z 134 (22%)                                        6%   18%

$\underline{C_{10}H_{14}}$, 134.1095 $(CH_3)_3$-phenyl-$CH_2$-,
phenyl-CO-OC$_4$H$_9$, etc                          17    16    73

$\underline{C_8H_6O_2}$, 134.0367 CH$_3$O-CO-phenyl-,
ext-ar(-O-CO-), $\underline{o}$-HO-phenyl-$CH_2$-CO-     7    20    54

$\underline{C_9H_{10}O}$, 134.0731 CH$_3$O-phenyl-CH$_2$CH$_2$-,
phenyl-$CH_2$-CO-$CH_2$-, cyc/unsatd ketones        9    13    55

$\underline{C_8H_8NO}$, 134.0605 CH$_3$NH-phenyl-CO-,
CH$_3$O-phenyl-CH=N-, CH$_3$-CO-N(phenyl)-,
phenyl-CO-CH=N-                                      5    19    56

$\underline{C_9H_{12}N}$, 134.0968 phenyl-N(C$_2$H$_5$)CH$_2$-,
C$_2$H$_5$(CH$_3$)$_2$-pyridyl etc                  4    20    56

$\underline{C_7H_6N_2O}$, 134.0478 HO-ext-arN$_2$,
ext-arO-NH-, -NH-phenyl-CO-N(-)-                    3    20    54

$\underline{C_6H_6N_4}$, 134.0589 ext-arN$_4$ (purines)   2    30    50

$\underline{C_7H_4NO_2}$, 134.0241 HO-(CH$_3$-)pyridyl-CO-,
-O-phenyl-CH=N-O-                                   3    12    56

$\underline{C_8H_{10}N_2}$, 134.0842 C$_2$H$_5$NH-ext-arN,

| m/z, comp | Substructure, neighbor | Prop | Abnd | Spcf |
|---|---|---|---|---|
| ar-amines | | 2 | 16 | 46 |

also $C_6H_4N_3O$, 134.0352 (-NHNH-pyridyl-CO-); $C_5H_2N_4O$, 134.0225 (HO-ext-$arN_4$); $C_6H_2N_2O_2$, 134.0114; $C_7H_4NS$, 134.0067 (benzothiazole)

| m/z 135 (26%) | | 10% | 23% | |
|---|---|---|---|---|
| $\underline{C}_{10}\underline{H}_{15}$, 135.1173 polyunsatd/cyc hc | | 13 | 20 | 79 |

| | | | | |
|---|---|---|---|---|
| $\underline{C}_8\underline{H}_7\underline{O}_2$, 135.0445 HO-($CH_3$-)phenyl-CO-, | | | | |
| $CH_3$O-CO-phenyl-, $CH_3$O-phenyl-CO- | | 9 | 19 | 56 |

| | | | | |
|---|---|---|---|---|
| $\underline{C}_9\underline{H}_{11}\underline{O}$, 135.0809 phenyl-OC($CH_3$)$_2$-, | | | | |
| ar/unsatd/cyc C=O/-O-/-OH | | 9 | 17 | 62 |

| | | | | |
|---|---|---|---|---|
| $\underline{C}_8\underline{H}_9\underline{N}O$, 135.0683 $H_2$N-CO-phenyl-$CH_2$-, | | | | |
| $HOCH_2$-($CH_3$-)pyridyl-$CH_2$-, | | | | |
| phenyl-CO-$NHCH_2$-, $H_2$N-($CH_3$O-)phenyl- | | 4 | 17 | 60 |

| | | | | |
|---|---|---|---|---|
| $\underline{C}_7\underline{H}_5\underline{N}O_2$, 135.0319 $O_2$N-phenyl-$CH_2$-, | | | | |
| $H_2$N-(HO-)phenyl-CO- | | 3 | 23 | 50 |

| | | | | |
|---|---|---|---|---|
| $\underline{C}_7\underline{H}_3\underline{O}_3$, 135.0082 -O-phenyl-CO-O-, | | | | |
| ar C=O/-O-/-OH | | 3 | 18 | 50 |

also $C_9H_{13}N$, 135.1047 (ar-amines); $C_5H_3N_4O$, 135.0304 (ext-$arN_3$-NH-CO- etc); $C_6H_5N_3O$, 135.0430 (ext-$arN_3$(C=O)); $C_8H_{11}N_2$, 135.0920 (-$CH_2$NH-phenyl-$NHCH_2$-, aminopyridines); $C_6H_3N_2O_2$, 135.0193 (ext-$arN_2$O(C=O)); $C_7H_7N_2O$, 135.0556 (CH$_3$NH-CO-pyridyl-, -N(-)-phenyl-NH-CO-); $C_4H_8Br$, 134.9809 (Br-($CH_2$)$_4$-); $C_5H_{11}O_4$, 135.0656 (sugars); $C_7H_9N_3$, 135.0794 ($arN_3$); $C_6H_7N_4$, 135.0667 (purines); $C_5H_5N_4$, 135.0541 (purine-NH-); $C_5H_5N_5$, 135.0541 (purine-NH-)

| m/z, comp | Substructure, neighbor | Prop | Abnd | Spcf |
|---|---|---|---|---|

m/z 136 (20%)                                                         6%   22%

$\underline{C_{10}H_{16}}$, 136.1251 polyunsatd/cyc hc              16   22   82

$\underline{C_9H_{12}O}$, 136.0887 cyc/unsatd/ar
  HO-C=O/-O-                                                        8    17   57

$\underline{C_8H_8O_2}$, 136.0524 phenyl-$CH_2$-CO-O-,
  ar/unsatd/cyc C=O/-O-/-OH                                        8    16   50

$\underline{C_7H_4O_3}$, 136.0160 HO-CO-(HO-)phenyl-,
  (HO-)$_2$phenyl-CO-, ext-arO(CO)-,
  cyc/substd/unsatd ketones                                        4    27   51

$\underline{C_7H_6NO_2}$, 136.0397 $H_2$N-(HO-)phenyl-CO-,
  $CH_3$-pyridone-CO-, $CH_3$O-pyridyl-CO-,
  $O_2$N-phenyl-$CH_2$-                                             4    16   48

$\underline{C_9H_{14}N}$, 136.1125 cyc/substd/unsatd
  amines                                                           3    22   67
also $C_8H_{10}NO$, 136.0761 ($CH_3$O-phenyl-N($CH_3$)-,
  $CH_3$O-phenyl-CH($NH_2$)-; $C_5H_4N_4O$, 136.0382
  (HO-ext-ar$N_4$); $C_8H_{12}N_2$, 136.0998 ($C_3H_7$NH-pyridyl-);
  $C_7H_6$NS, 136.0223 (ext-arNS); $C_8H_8$S, 136.0350
  (ext-arS); $C_7H_{10}N_3$, 136.0872
  (($CH_3$)$_2$-pyrimidinyl-NHCH$_2$-); $C_6H_8N_4$, 136.0746
  ($H_2$N-triazinyl-R); $C_5H_6N_5$, 136.0619 ($H_2$N-purines)

m/z 137 (22%)                                                        6%   22%

$\underline{C_{10}H_{17}}$, 137.1329 decahydronaphthyl              15   15   85

$\underline{C_8H_9O_2}$, 137.0602 $CH_3$O-phenyl-CH(OH)-,
  $CH_3$-phenyl-CO-O-                                              9    25   55

# MASS SPECTRAL CORRELATIONS

| m/z, comp | Substructure, neighbor | Prop | Abnd | Spcf |
|---|---|---|---|---|

$\underline{C_9H_{13}O}$, 137.0966 cyc/unsatd/substd
  C=O/-O-/-OH

|  |  | 11 | 16 | 68 |
|---|---|---|---|---|

$\underline{C_7H_5O_3}$, 137.0238 (HO-)$_2$phenyl-CO-,
  HO-(-O-)phenyl-CH(OH)-

|  |  | 6 | 26 | 60 |
|---|---|---|---|---|

$\underline{C_7H_7NO_2}$, 137.0475 $H_2$N-phenyl-CO-O-,
  phenyl-NH-CO-O-, HO-CO-phenyl-NH-

|  |  | 4 | 25 | 59 |
|---|---|---|---|---|

$\underline{C_8H_9S}$, 137.0428 $CH_3$-phenyl-$SCH_2$-

|  |  | 2 | 31 | 83 |
|---|---|---|---|---|

$\underline{C_8H_{11}NO}$, 137.0839 $C_2H_5$O-($CH_3$-)-pyridyl-,
  $CH_3$O-($H_2NCH_2$-)phenyl-

|  |  | 2 | 19 | 50 |
|---|---|---|---|---|

also $C_9H_{15}N$, 137.1203 (cyc/unsatd amines); $C_7H_4NCl$,
137.0031 (Cl-ext-arN-, Cl-phenyl-CH=N-);
$C_4H_{10}O_3P$, 137.0367 (cyc-$CH_2$-CH(P(=O)($OCH_3$)$_2$)-,
($C_2H_5$O)$_2$P(=O)-); $C_6H_3NO_3$, 137.0112
(HO-ar-(-O-)-NH-CO-, HO-CO-ar-NH-CO-); $C_7H_9N_2O$,
137.0713

| m/z 138 (17%) |  | 4% | 20% |  |
|---|---|---|---|---|

$\underline{C_{10}H_{18}}$, 138.1408 cyc/unsatd hc

|  |  | 15 | 12 | 84 |
|---|---|---|---|---|

$\underline{C_8H_{10}O_2}$, 138.0680 cyc/substd/unsatd
  C=O/-O-, (-CO-$C_6H_{12}$-CO-,
  cycR-O-CO-$CH_3$),
  $CH_3$O-(HO-)($CH_3$-)-phenyl-

|  |  | 7 | 21 | 56 |
|---|---|---|---|---|

$\underline{C_9H_{14}O}$, 138.1044 cyc/substd/unsatd
  ketones

|  |  | 8 | 17 | 57 |
|---|---|---|---|---|

$\underline{C_7H_6O_3}$, 138.0316 HO-CO-phenyl-O-,
  -O-phenyl-O-CO-, etc

|  |  | 6 | 22 | 50 |
|---|---|---|---|---|

| m/z, comp | Substructure, neighbor | Prop | Abnd | Spcf |
|---|---|---|---|---|

$\underline{C_9H_{16}N}$, 138.1281 cyc/substd/unsatd
amines, $NC-C_8H_{16}-$      6    19    61

$\underline{C_6H_6N_2O_2}$, 138.0427 $O_2N$-phenyl-NH-,
HO-CO-pyridyl-NH-, pyrimidione-$C_2H_4-$    2    38    48

$\underline{C_8H_{12}NO}$, 138.0917 cyc/substd/unsatd
amines/-O-, -furyl-N-      4    21    47

$\underline{C_7H_8NO_2}$, 138.0554 cyc/ar/substd/unsatd
amines/-O-/-OH/C=O,
$CH_3O-(HO-)(CH_3-)$-pyridyl, ketolactam    3    18    50
also $C_7H_{10}N_2O$, 138.0791 ($C_2H_5O-(CH_3-)$-pyrimidinyl-)

| m/z 139 (22%) | 6% | 21% |
|---|---|---|

$\underline{C_{10}H_{19}}$, 139.1486 cyc/unsatd hc      22    15    87

$\underline{C_9H_{15}O}$, 139.1122 2-or 3-$C_6H_{13}-$
cycloalkanone, -$C_8H_{16}$-CO-, cyc/substd
ketones; and $\underline{C_{10}H_3O}$, 139.0184
ext-ar(C=O)      9    18    66

$\underline{C_7H_4OCl}$, 138.9950 Cl-phenyl-CO-      2    88    74

$\underline{C_7H_7O_3}$, 139.0394 furanoates, ar
-CO-O-/-O-      5    23    64

$\underline{C_8H_{11}O_2}$, 139.0758 ar(C=O)-O-, cyc/unsatd
C=O/-O-/-OH      5    16    58

$\underline{C_{10}H_5N}$, 139.0421 ext-arN
$\underline{C_9H_{17}N}$, 139.1359 cyc/substd amines    4    12    63

$\underline{C_6H_3O_4}$, 139.0031 HO-CO-furyl-$CH_2O-$,

| m/z, comp | Substructure, neighbor | Prop | Abnd | Spcf |
|---|---|---|---|---|
| HO-CO-furyl-CO-, ar-O-/C=O | | 2 | 24 | 50 |

also $C_7H_9NO_2$, 139.0632 (arN-CO-O-, ar-NH-CO-O);
$C_8H_{13}NO$, 139.0996; $C_6H_7N_2O_2$, 139.0506; $C_8H_{11}S$,
139.0584 ($C_4H_9$-thiophenyl-); $C_7H_7OS$, 139.0220
(HO-phenyl-S-$CH_2$-) $C_8H_8Cl$, 139.0314
(Cl-phenyl-($C_2H_5$)-); $C_9H_3N_2$, 139.0295 (ext-$arN_2$);
$C_6H_5NO_3$, 139.0268 ($O_2N$-phenyl-O-)

| m/z 140 (17%) | | | 4% | 16% |
|---|---|---|---|---|
| $\underline{C_{10}H_{20}}$, 140.1564 H-$C_{10}H_{20}$-Y*; and | | | | |
| $\underline{C_{11}H_8}$, 140.0626 naphthyl-$CH_2$- | | 16 | 8 | 84 |
| $\underline{C_9H_{18}N}$, 140.1438 cyc/substd/unsatd | | | | |
| amines; and $\underline{C_{10}H_6N}$, 140.0499 | | | | |
| $CH_3$-quinolinyl, benzoazepinyl | | 9 | 12 | 71 |
| $\underline{C_9H_{16}O}$, 140.1200 cyc/substd/unsatd | | | | |
| -CO-/-O-/-OH; and $\underline{C_{10}H_4O}$, 140.0262 | | | | |
| ext-ar(C=O) | | 8 | 11 | 62 |
| $\underline{C_8H_{12}O_2}$, 140.0836 cyc/substd/unsatd | | | | |
| C=O/-O- etc | | 5 | 14 | 55 |
| $\underline{C_8H_{14}NO}$, 140.1074 cyc-$C_6H_{11}$-N(-CO-$CH_3$)-, | | | | |
| cyc/unsatd amides, oximes, HO-amines, | | | | |
| isocyanates | | 4 | 17 | 59 |
| $\underline{C_7H_8O_3}$, 140.0473 cyc/substd/unsatd | | | | |
| -O-/C=O/-OH ($CH_3O$-pyrone-$CH_2$-) | | 4 | 20 | 60 |

also $C_7H_{10}NO_2$, 140.0710 (ar-NH-CO-$OC_2H_5$); $C_7H_{10}NS$,
140.0536 (ar-S-R, arNS); $C_9H_4N_2$, 140.0373 (ext-$arN_2$);
$C_6H_8N_2O_2$, 140.0584 (unsatd amides, carbamates);
$C_8H_{12}S$, 140.0662 (arS-R); $C_6H_4O_4$, 140.0109

| m/z, comp | Substructure, neighbor | Prop | Abnd | Spcf |
|---|---|---|---|---|

<u>m/z 141 (23%)</u>          6%   19%

$\underline{C_{11}H_9}$, 141.0704 naphthyl-$CH_2$-, ext-ar hc   24   18   90

$\underline{C_9H_{17}O}$, 141.1278 $C_8H_{17}$-CO-,
-CH(OH)-$(CH_2)_8$-; and $\underline{C_{10}H_5O}$, 141.0340
-O-phenyl-ar-O-, ext-ar(C=O) etc    8   13   71

$\underline{C_8H_{13}O_2}$, 141.0915 $C_5H_{11}$-CO-$CH_2$-CO-,
$CH_3$O-CO-cyclohexyl(-)-,
$CH_3$O-CO-$(CH_2)_3$C≡C$CH_2$-,
$C_4H_9$C(O$CH_3$)=CH-CO-       5   14   60

$\underline{C_8H_{13}S}$, 141.0741 cyc/substd/unsatd
sulfides                   2   37   56

$\underline{C_7H_9O_3}$, 141.0551 ar/cyc/unsatd
C=O/-O-/-OH             3   19   54

$\underline{C_7H_{11}NO_2}$, 141.0788 R-ar(-CO-NH-CO-)-,
succinimides           2   30   60
also $C_{10}H_7N$, 141.0577 (ext-arN); $C_9H_5N_2$, 141.0451
(ext-arN-N- etc); $C_6H_5O_4$, 141.0187
(cyc-CH(-CO-O-)-CH(-CO-O-)-); $C_7H_{11}NS$, 141.0614
(arN-S-); $C_7H_6OCl$, 141.0106 (HO-($CH_3$-)(Cl-)phenyl-,
HO-(Cl-)phenyl-$CH_2$-); $C_6H_5O_2S$, 141.0013
(phenyl-$SO_2$-, HO-phenyl-SO-); $C_6H_9N_2O_2$, 141.0662
(($CH_3$)$_2$NC(=CH-CO-$CH_3$)-N(-)-); $CH_2I$, 140.9199
(iodo-$CH_2$-)

<u>m/z 142 (18%)</u>          4%   19%

$\underline{C_{11}H_{10}}$, 142.0782 naphthyl-$CH_2$-, ext-ar hc   19   12   85

| m/z, comp | Substructure, neighbor | Prop | Abnd | Spcf |
|---|---|---|---|---|

$\underline{C_{10}H_8N}$, 142.0656 ext-arN ($CH_3$quinolines),
naphthyl-NH-, -phenyl-cycN $\qquad$ 7 $\quad$ 23 $\quad$ 68

$\underline{C_8H_{14}O_2}$, 142.0993 $CH_3O\text{-}CO\text{-}(CH_2)_6\text{-}$,
$(CH_3O)_2$-cyclohexyl- $\qquad$ 4 $\quad$ 14 $\quad$ 54

$\underline{C_8H_{16}NO}$, 142.1230 $-(CH_2)_7\text{-}NH\text{-}CO\text{-}$,
$CH_3\text{-}CO\text{-}N(C_5H_{11})CH_2\text{-}$,
HO-(R)-pyrollidinyl-; and
$\underline{C_9H_4NO}$, 142.0291 ext-arN(C=O)
(quinolinonyl), ext-arN-O- $\qquad$ 4 $\quad$ 21 $\quad$ 55

also $C_9H_6N_2$, 142.0529 (ext-arN$_2$, ext-arN-CH=N-);
$C_8H_{18}N_2$, 142.1468 ($C_8H_{17}\text{-}N(\text{-})\text{-}N(\text{-})\text{-}$); $C_9H_{18}O$,
142.1357 ($C_7H_{15}\text{-}CO\text{-}CH_2\text{-}$); $C_{10}H_6O$, 142.0418 (ext-arO);
$C_7H_{10}O_3$, 142.0629 ($CH_3O\text{-}CO\text{-}C_4H_8\text{-}CO\text{-}$); $C_7H_{12}NO_2$,
142.0866 ($CH_3\text{-}CO\text{-}N(C_4H_9)\text{-}CO\text{-}$); $C_6H_6O_2S$, 142.0091
(HO-phenyl-SO-, phenyl-SO$_2$-); $C_7H_{12}NS$, 142.0692
(arN-S-); $C_6H_{10}N_2S$, 142.0566 (R-imidazolyl-S-);
$C_7H_7OCl$, 142.0184 (Cl-($CH_3$-)phenyl-O-); $C_6H_8NOS$,
142.0329 (HO-($CH_3$-)-pyrrolyl-S-$CH_2$-); $C_6H_{10}N_2O_2$,
142.0740 (cyc-$CH_2CH(OH)CH(CH_3)N(\text{-})\text{-}CO\text{-}N(CH_3)\text{-}$);
$C_6H_6O_4$, 142.0265 ($C_2H_5O\text{-}CO\text{-}C(\text{-})_2CH_2\text{-}CO\text{-}O\text{-}$)

| m/z 143 (23%) | | | 6% | 19% |
|---|---|---|---|---|

$\underline{C_{11}H_{11}}$, 143.0860 substd indenes,
tetralins etc $\qquad$ 18 $\quad$ 15 $\quad$ 82

$\underline{C_{10}H_9N}$, 143.0734 naphthyl-NH-,
quinolinyl-$CH_2$-, phenyl-pyrrolyl etc $\qquad$ 7 $\quad$ 18 $\quad$ 66

$\underline{C_9H_7N_2}$, 143.0607 ext-arN$_2$
($CH_3$-quinoxalinyl-), NC-ext-arN,
phenyl-C(-CN)=CH-NH- $\qquad$ 6 $\quad$ 19 $\quad$ 69

| m/z, comp | Substructure, neighbor | Prop | Abnd | Spcf |
|---|---|---|---|---|

$\underline{C_9H_{19}O}$, 143.1435  $(C_4H_9)_2C(OH)-$,
 $C_7H_{15}CH(OCH_3)-$                                                  6    14    68

$\underline{C_7H_{11}O_3}$, 143.0707  $CH_3O-CO-C_4H_8-CO-$,
 $CH_3O-CO-CH_2CH_2-CH=C(OCH_3)-$                                       5    16    63

$\underline{C_8H_{15}O_2}$, 143.1071  $CH_3O-CO-(CH_2)_6-$,
 $-(CH_2)_5CH(-)O-CO-$,
 $C_2H_5O-CO-CH_2CH_2C(-)(C_2H_5)-$; and
 $\underline{C_9H_3O_2}$, 143.0133  ext-ar-$(C=O)_2-$,
 ext-ar$(C=O)_2$                                                       6    13    61
also $C_9H_5NO$, 143.0370 (phenyl-arN(C=O)); $C_6H_7O_4$,
143.0343 ($-CH(CH_3)CH(-CO-OCH_3)-CO-O-$); $C_6H_{11}N_2O_2$,
143.0818 ($-CH_2CH_2C(-CH_2-)=N-NH-CO-OCH_3$)

| m/z 144 (18%) | | 5% | 22% |
|---|---|---|---|

$\underline{C_{11}H_{12}}$, 144.0938  phenyl-$C_5H_7(-)-$,
 $C_6H_{11}$-phenyl-$CH_2-$,  ext-ar                                   15   13    77

$\underline{C_{10}H_{10}N}$, 144.0812  ext-arN
 $((CH_3)_2$-indolinyl-), phenyl-arN                                   7    39    67

$\underline{C_9H_6NO}$, 144.0448  indole-CO-,
 phenyl-isoxazolyl-                                                    6    31    58

$\underline{C_{10}H_8O}$, 144.0575  naphthyl-O-                        8    20    63

$\underline{C_8H_{16}O_2}$, 144.1149  $C_2H_5O-CO-C(C_2H_5)_2-$;
 and $\underline{C_9H_2O_2}$, 144.0211  ext-ar$(C=O)_2$               5    17    60

$\underline{C_9H_8N_2}$, 144.0686  $C_2H_5$-benzimidazolyl-,
 phenyl-C(CN)=CH-NH-                                                   4    21    58
also $C_8H_4N_2O$, 144.0322 (HO-quinoxalinyl-); $C_6H_8O_4$,

MASS SPECTRAL CORRELATIONS

m/z, comp    Substructure, neighbor        Prop Abnd Spcf
144.0422 $(-C(-O-CO-C_2H_5)_2)$; $C_7H_{14}NO_2$, 144.0123
$(HO-CO-(CH_2)_5-CH(NH_2)-)$; $C_6H_{10}NO_3$, 144.0659
$(-CH_2CH_2CH(-CO-OCH_3)-NH-CO-)$

m/z 145 (23%)                                    9%   20%

$\underline{C_{11}H_{13}}$, 145.1017 phenyl-unsatd R,
benzo-cycR                                       17   23   80

$\underline{C_{10}H_9O}$, 145.0653 $CH_3$-phenyl-CH=CH-CO-,
ext-ar-O-                                         9   18   64

$\underline{C_6H_9O_4}$, 145.0500 $CH_3O-CO-CH_2-CO-CH_2CH(OH)-$,
cyc-CH($-CO-OCH_3$)-C(-)($-CO-OCH_3$)-            3   25   64

$\underline{C_9H_7NO}$, 145.0526 quinoline-O-, ext-arN-OH,
ext-ar(-NH-CO-)                                   5   13   61

$\underline{C_{10}H_{11}N}$, 145.0890 R-ext-arN, R-indolinyl-  4   17   61

$\underline{C_8H_5N_2O}$, 145.0400 ext-arN(-N-CO-),
HO-ext-arN$_2$-, phenyl-arN$_2$O                  3   15   68

$\underline{C_9H_9N_2}$, 145.0764 $C_2H_5$-benzimidazolyl-,
phenylpyrazole                                    2   21   60
also $C_8H_{17}O_2$, 145.1227 $(C_5H_{11}OCH(OC_2H_5)-)$; $C_9H_5O_2$,
145.0289 (-O-ext-ar-CO-); $C_7H_{17}OSi$, 145.0947
$(CH_3)_3SiOCH(C_3H_7)-$, $C_5H_{11}OSi(CH_3)_2-)$; $C_7H_3N_3O$,
145.0274 (ext-arN$_2$-NO); $C_6H_{13}O_2Si$, 145.0584
$(C_3H_7-CO-O-Si(CH_3)_2-)$; $C_6H_{11}NO_3$, 145.0737
$(HO-CO-CH(C_3H_7)-NH-CO-)$; $C_6H_3Cl_2$, 144.9611
($Cl_2$-phenyl-)

| m/z, comp | Substructure, neighbor | Prop | Abnd | Spcf |
|---|---|---|---|---|

m/z 146 (19%)                                              5%    20%

$C_{10}H_{10}O$, 146.0731 $(CH_3)_3$phenyl-CO-,
  phenyl-CH=CH-CO-$CH_2$-,
  benzocyclohexanones,
  cyc-C($C_2H_5$)(phenyl)-CO-                        8     24     58

$C_{10}H_{12}N$, 146.0968 benzo-cycN,
  -CH(-$CH_2$phenyl)NHCH($CH_3$)-                  6     25     70

$C_{11}H_{14}$, 146.1095 ar/cyc/unsatd hc          12     12     76

$C_9H_8NO$, 146.0605 HO-benzopyrrolyl-$CH_2$-
  etc, phenyl-$CH_2$CH(-)NH-CO-                       5     29     60

$C_9H_6O_2$, 146.0367 HO-phenyl-CH=CH-CO-,
  phenyl-CH=C(-CO-OH)-, ext-ar(C=O)$_2$,
  cyc-CO-C(-)(phenyl)-CO-                             6     21     65

$C_8H_6N_2O$, 146.0478 ext-arN$_2$-O-,
  ext-arN$_2$(C=O)                                    2     33     50

$C_8H_4NO_2$, 146.0241 o-CO-phenyl-NH-CO-,
  ext-arN(C=O)$_2$, HO-CO-benzopyrrolyl,
  HO-ext-arN(C=O)                                     2     24     61

$C_9H_{10}N_2$, 146.0842 $(CH_3)_2$N-phenyl-CH=N-,
  benzo-cycN$_2$-, $H_2$NCH$_2$-ext-arN           3     14     52
also $C_8H_8N_3$, 146.0716 (ext-arN$_3$); $C_6H_{10}O_4$, 146.0578
  ((HO-CO)$_2$-C($C_3H_7$)-)

MASS SPECTRAL CORRELATIONS

| m/z, comp | Substructure, neighbor | Prop | Abnd | Spcf |
|---|---|---|---|---|

| m/z 147 (25%) | | 12% | 23% | |
|---|---|---|---|---|

$\underline{C}_{11}\underline{H}_{15}$, 147.1173 $(CH_3)_2$phenyl$C(CH_3)_2-$,
$(CH_3)_5$phenyl etc
                                                15   23   76

$\underline{C}_9\underline{H}_7\underline{O}_2$, 147.0445 phenyl$-CO-CH_2-CO-$,
HO$-$phenyl$-CO-CH=CH-$, $CH_3O-$benzofuryl$-$   8   19   65

$\underline{C}_{10}\underline{H}_{11}\underline{O}$, 147.0809 $(CH_3)_3$-phenyl$-CO-$,
$CH_3O-$phenyl$-CH_2CH=CH-$,
phenyl$-CH_2CH(-)CH(OCH_3)-$,
substd/cyc/unsatd ketones                   8   18   63

$\underline{C}_8\underline{H}_7\underline{N}_2\underline{O}$, 147.0556 oxazolyl$-NHCH_2-$,
$H_2N-NH-(CH_3-)$phenyl$-CO-$, $CH_3O-$ext$-arN_2$   2   32   60

$\underline{C}_9\underline{H}_9\underline{N}\underline{O}$, 147.0683 $(CH_3)_2$phenyl$-NH-CO-$,
$-$phenyl$-CO-CH_2CH(NH_2)-$,
cyc$-CH(CH_3)-N($phenyl$)-CO-$             3   20   54

also $C_{10}H_{13}N$, 147.1047 $((C_2H_5)_2N-$phenyl$-)$; $C_5H_{15}Si_2O$,
147.0400 (rearr; $\geq 2$ $(CH_3)_3SiO-$groups); $C_8H_5NO_2$,
147.0319 (ext$-ar(-CO-N-CO-))$; $C_9H_{11}N_2$, 147.0920
$((CH_3)_2N-CH=N-$phenyl$-$, ext$-arN_2$ amines); $C_8H_3O_3$,
147.0082 ($-$o$-$benzo$-$cyc$-CH_2O-CO-)$; $C_7H_3N_2O_2$, 147.0193
(ext$-arN_2(C=O)-CO-)$; $C_9H_7S$, 147.0271
(benzothiophenyl$-CH_2-)$; $C_7H_5N_3O$, 147.0430
(ext$-arN_3(C=O))$; $C_7H_7N_4$, 147.0667 (ext$-arN_4$)

| m/z 148 (21%) | | 5% | 18% | |
|---|---|---|---|---|

$\underline{C}_{11}\underline{H}_{16}$, 148.1251 $(CH_3)_4$-phenyl$-CH_2-$ etc   13   15   74

$\underline{C}_9\underline{H}_8\underline{O}_2$, 148.0524 $CH_3O-CO-CH_2-$phenyl$-$,

| m/z, comp | Substructure, neighbor | Prop | Abnd | Spcf |
|---|---|---|---|---|
| -phenyl-CH$_2$-CO-OCH$_2$-, | | | | |
| HO-phenyl-CH=CH-CO-O- | | 7 | 21 | 55 |

$\underline{C}_9\underline{H}_{10}\underline{NO}$, 148.0761 phenyl-CO-NHCH(CH$_3$)-,
(CH$_3$)$_2$N-phenyl-CH(OH)-

|  |  | 5 | 21 | 57 |
|---|---|---|---|---|

$\underline{C}_{10}\underline{H}_{12}\underline{O}$, 148.0887 phenyl-CH$_2$CH$_2$-CO-CH$_2$-,
cyc/unsatd/ar C=O/-O-/-OH

|  |  | 6 | 15 | 62 |
|---|---|---|---|---|

$\underline{C}_8\underline{H}_4\underline{O}_3$, 148.0160 HO-CO-phenyl-CO-,
-CH$_2$-(HO-)phenyl-CO-O-

|  |  | 3 | 26 | 49 |
|---|---|---|---|---|

$\underline{C}_8\underline{H}_6\underline{NO}_2$, 148.0397 pyridyl-CO-CH$_2$-CO-,
HO-CO-CH$_2$-pyridyl-,
HO-N=CH-CH$_2$-phenyl-O-

|  |  | 3 | 21 | 46 |
|---|---|---|---|---|

also C$_8$H$_8$N$_2$O, 148.0635 (H$_2$N-NH-(CH$_3$-)phenyl-CO-);
C$_{10}$H$_{14}$N, 148.1125 (arN, phenyl amines); C$_9$H$_{12}$N$_2$,
148.0998 (phenyl diamines); C$_7$H$_4$N$_2$O$_2$, 148.0271
(O$_2$N-phenyl-CH=N-); C$_6$H$_2$N$_3$O$_2$, 148.0145
(ext-arN$_3$(C=O)-O-); C$_6$H$_4$N$_4$O, 148.0382 (ext-arN$_3$-NO);
C$_7$H$_8$N$_4$, 148.0746 (ext-arN$_4$, (CH$_3$)$_2$N-ext-arN$_3$);
C$_6$H$_6$N$_5$, 148.0620 (purine-NH-CH$_2$-); C$_8$H$_{10}$N$_3$,
148.0872 (phenyl-NHC(=N-NHCH$_3$)-); C$_6$F$_4$, 147.9936
(tetrafluorophenyl)

| m/z 149 (26%) | | 9% | 19% | |
|---|---|---|---|---|

$\underline{C}_{11}\underline{H}_{17}$, 149.1329 polyunsatd/cyc hc

|  | | 12 | 18 | 81 |
|---|---|---|---|---|

$\underline{C}_9\underline{H}_9\underline{O}_2$, 149.0602 phenyl-CH(OH)CH$_2$-CO-,
ar C=O/-O-/-OH

|  | | 8 | 16 | 64 |
|---|---|---|---|---|

$\underline{C}_8\underline{H}_5\underline{O}_3$, 149.0238 phthalates,
HO-CO-phenyl-CO-

|  | | 5 | 21 | 59 |
|---|---|---|---|---|

| m/z, comp | Substructure, neighbor | Prop | Abnd | Spcf |
|-----------|------------------------|------|------|------|

$\underline{C_{10}H_{13}O}$, 149.0966 cyc/substd/unsatd/ar
  C=O/-OH/-O-                  7    15    67

$\underline{C_8H_7NO_2}$, 149.0475 ON-CH$_2$-phenyl-CO-,
  arN-CH$_2$-CO-O-               3    18    51

also $C_9H_{11}NO$, 149.0839 (ar -O-/amines); $C_{10}H_{15}N$,
  149.1203 (arN); $C_7H_5N_2O_2$, 149.0349 (ar); $C_9H_{13}N_2$,
  149.1077 ($C_2H_5$NH-phenyl-NHCH$_2$-, pyrazines);
  $C_6H_3N_3O_2$, 149.0223 (ext-arN$_3$(C=O)$_2$); $C_8H_7$NS,
  149.0301 (benzothiazole-CH$_2$-)

| m/z 150 (19%) | | 5% | 20% | |
|---------------|--|----|-----|--|

$\underline{C_{12}H_6}$, 150.0469 ext-ar hc            17    13    84

$\underline{C_9H_{10}O_2}$, 150.0680 phenyl-CH(-CO-OCH$_3$)-,
  cyc/unsatd/ar C=O/-OH/-O-        7    14    59

$\underline{C_7H_4NO_3}$, 150.0190 O$_2$N-phenyl-CO-,
  -O-phenyl-O-CO-NH-             2    60    68

$\underline{C_{10}H_{14}O}$, 150.1044 cyc/unsatd ketones,
  phenyl -O-/-OH                  5    20    57

$\underline{C_8H_6O_3}$, 150.0316 phenyl
  -CO-O-/-OH/CH$_3$-/-CO-OH/-O-     4    20    53

$\underline{C_8H_8NO_2}$, 150.0554 CH$_3$O-CO-NH-phenyl-,
  ar/cyc/unsatd
  amines/-CO-O-/C=O/-O-/-CO-NH-    4    24    58

also $C_{10}H_{16}N$, 150.1281 (cyc/substd amines); $C_7H_6N_2O_2$,
  150.0427 (HO-CO-ext-arN$_2$, ON-phenyl-NH-CO-);
  $C_9H_{12}NO$, 150.0918 ((CH$_3$)$_2$N-(HO-)(CH$_3$-)phenyl-);
  $C_5H_2N_4O_2$, 150.0174 (O$_2$N-ext-arN$_3$); $C_7H_6N_2$S,

m/z, comp    Substructure, neighbor      Prop Abnd Spcf
150.0253 (ext-arNS-NH-);  $C_3F_6$,  149.9904;  $C_6H_2N_2O_3$,
150.0063 ($O_2$N-(-N-CO-)$_2$-ar-)


Note:   The computer-aided correlations were carried
        out only for data from m/z 29 to m/z 150,
        inclusive.


m/z 151
_____

$C_2Cl_2F_3$, 150.9329

$C_9H_{11}O_2$, 151.0758 (HO)$_2$phenyl-C(CH$_3$)$_2$-

$C_8H_7O_3$, 151.0720 (CH$_3$O)$_2$phenyl-CH$_2$,
    CH$_3$O-(HO)-phenyl-CO-,  CH$_3$O-CO-phenyl-O-
  also IC≡C-, CH$_2$-CH-(CH$_3$-)(Cl-)phenyl-, monoterpenones,
    $C_{12}H_7$


m/z 152
_____

$C_{12}H_8$, 152.0626 -phenyl-phenyl-, phenyl-CH(-)-phenyl
  also CH$_3$O-C$_7$H$_5$O$_2$, O$_2$N-(HO-)phenyl-CH$_2$-, CH$_2$=CH-CO-N
    (cyclohexyl)-, Cl-benzoxazolyl-, aporphine alkaloids


m/z 153
_____

$C_9H_{10}Cl$, 153.0471 Cl-phenyl dvts

$C_{12}H_9$, 153.0704 phenyl-phenyl-, naphthyl-CH=CH-
  also BrC$_2$H$_4$-CO-O-, (CH$_3$O)$_2$-phenyl-O-,
    CH$_3$O-phenyl-CO-O-, thiophenyl-CO-CH$_2$-CO-


m/z 154
_____

H-CO-NH-(Cl-)phenyl-, (phenyl)$_2$SiCl$_2$,

| m/z, comp | Substructure, neighbor | Prop | Abnd | Spcf |
|---|---|---|---|---|

$CF_3-CO-N(C_2H_5)CH_2-$ , $HO-(Cl-)(R-)$phenyl-

---

m/z 155

---

$Cl-(R-)$phenyl-$O-$ , $Br$-phenyl-, $HO-(Cl-)$phenyl-$CO-$ ,
$(C_2H_5O-)_2$ , $P(=O)O-$ , $C_4H_9O-P(=O)(-O-)_2$ , $C_3H_3{}^{81}BrCl$ ,
$C_5F_5$ , $C_3H_2BrF_2$ , phenyl-$(CH_3O-)PO-$ , naphthyl-$CO-$ ,
$CH_3$-phenyl-$SO_2-$ , $C_4H_9O-CO-C_2H_2-CO-$ , (pyridyl)$_2-$ ,
$CH_3$-naphthyl-$CH_2-$ , $C_{11}H_{23}$

---

m/z 156

---

$C_8H_{17}N(CH_3)CH_2-$ , etc, $CH_3$-quinolinyl-$CH_2-$ ,
quinolinyl-$CH_2CH_2-$ , $Br$-pyridyl-, $(C_4H_9)_2N-CO-$

---

m/z 157

---

$C_3H_5{}^{81}BrCl$ , 156.9244; $C_3H_4BrF_2$ , 156.9465; $(C_3H_7)_3Si-$ ,
$CH_3$-phenyl-pyrazole-

---

m/z 158

---

$\underline{C_{11}H_{12}N}$, 158.0969 $(CH_3)_2$-indole-$CH_2-$

$\underline{C_{12}H_{14}}$, 158.1095 -phenyl-cyclohexyl-

---

m/z 159

---

$\underline{C_7H_5Cl_2}$, 158.9768 $Cl_2$-phenyl-$CH_2-$ , phenyl-$CCl_2-$
also $C_{12}H_{15}$, $C_6H_8Br$, $CHFI$

---

m/z 160

---

$C_2BrF_3$ , $CH_3O$-indole-$CH_2-$ , $-Cl_2$-phenyl-$O-$

m/z, comp     Substructure, neighbor     Prop Abnd Spcf

m/z 161

$C_{12}H_{17}$, 161.1329; $CH_3$-benzothiophene-$CH_2$-, $BrC_6H_{10}$-, $C_3H_5$-, CO-N(phenyl)-

m/z 162

$\underline{C_6H_4OCl_2}$, 161.9639 $Cl_2$-phenyl-O-
also phenyl-N($C_4H_9$)-$CH_2$-, phenyl-N(-CO-$CH_3$)CH($CH_3$)-

m/z 163

$C_3Cl_2F_3$, 162.9329; $C_{11}H_{15}O$, 163.1122; $C_{10}H_{11}O_2$, $C_9H_7O_3$, $C_{12}H_{19}$, $C^{81}BrCl_2$, $(C_2H_5O)_3Si$-, $C_4H_4ClF_4$, $C_3OClF_4$, $C_{10}H_{13}ON$

m/z 164

$C_6H_{11}$-$C_6H_9$-, $C_3H_5$-($CH_3O$-)phenyl-O-, H-CO-($CH_3O$-)phenyl-O-, -fluorene-, berbines

m/z 165

$\underline{C_{13}H_9}$, 165.0704 phenyl-CH(-)-phenyl-, -($CH_3$-)phenyl-phenyl-
also $C_{10}H_{13}O_2$, $C_9H_9O_3$, aporphine alkaloids, $BrCH_2$-CO-OCH($CH_3$)-, Cl-phenyl-CH=CH-CO-, Cl-benzofuran-$CH_2$-, $(C_2H_5O)_2P(=O)C_2H_4$-, $C_5Cl_3$

m/z 166

$\underline{C_{13}H_{10}}$, 166.0782 (phenyl)$_2$-C(-)-
also $C_2Cl_3{}^{37}Cl$, -($C_3H_5$-)(Cl-)phenyl-O-, carbazole-, $O_2N$-(HO-)phenyl-CO-, phenyl(-CO-O-)$_2$,

m/z, comp    Substructure, neighbor        Prop Abnd Spcf
  Cl-phenyl-N($C_3H_5$)-

m/z 167
─────────────────────────────────────────────

$\underline{C}_2\underline{Cl}_3\underline{F}_2$, 166.9034

$\underline{C}_{13}\underline{H}_{11}$, 167.0860 (phenyl)$_2$-CH-, phenyl-CH$_2$-phenyl-,
  acenaphthenes
also Cl-phenyl-N(-CH$_2$CH=CH$_2$)-, $C_{10}H_{12}Cl$,
  Cl-(HO-)phenyl-, $C_3H_4$-, phenyl(-CO-O-)$_2$

m/z 168
─────────────────────────────────────────────

-phenyl-O-phenyl-, (phenyl)$_2$-N-, $C_9H_9OCl$,
  O$_2$N-phenyl-O-, $C_8H_8O_4$, phenyl-NH-CO-, (phenyl)$_2$CH-

m/z 169
─────────────────────────────────────────────

$\underline{C}_{12}\underline{H}_9\underline{O}$, 169.0653 phenyl-phenyl-O-, phenyl-O-phenyl-,
  HO-(phenyl)$_2$-
also $C_7H_6Br$, 168.9653; $C_9H_{10}OCl$, 169.0420; (phenyl)$_2$N-,
  Cl-phenyl-Si(CH$_3$)$_2$-, naphthyl-C$_3$-,
  Cl$_2$C$_2$H$_3$-CO-OCH(CH$_3$)-, $C_3H_5$O-CO-C$_4$H$_8$-CO-,
  Cl-terpenoles, $C_3F_7$, ClCH$_2$-CO-N(phenyl)-, $C_{12}H_{25}$,
  -(Br-)(HO-)phenyl-

m/z 170
─────────────────────────────────────────────

$\underline{C}_{12}\underline{H}_{10}\underline{O}$, 170.0731 phenyl-phenyl-O-
also (C$_5$H$_{11}$)$_2$NCH$_2$-, Br-phenyl-NH-, (pyridyl)$_2$N-,
  C$_3$H$_7$-(Cl-), phenyl-O-

m/z 171
─────────────────────────────────────────────

$\underline{C}_8\underline{H}_5\underline{Cl}_2$, 170.9768 -(Cl$_2$-phenyl)-CH(-)CH$_2$-,

| m/z, comp | Substructure, neighbor | Prop | Abnd | Spcf |
|---|---|---|---|---|

$Cl_2$-phenyl-CH=CH-

$\underline{C_{12}H_{11}O}$, 171.0809 phenyl-phenyl-O-
also -CO-$C_7H_{14}$-CO-O-, F-phenyl-phenyl-, $C_5ClF_4$,
 $CH_3$O-phenyl-$SO_2$-

## m/z 172

$\underline{C_6H_5OBr}$, 171.9524 Br-phenyl-O-

$\underline{C_8H_6Cl_2}$, 171.9846 -$Cl_2$-phenyl-$C_2H_3$-

## m/z 173

$\underline{C_8H_7Cl_2}$, 172.9924 $CH_3$-($Cl_2$-)phenyl-$CH_2$-,
 $Cl_2$-phenyl-CH($CH_3$)-, $Cl_2$-phenyl-$CH_2CH_2$-
also $Cl_2$-phenyl-CO-, $C_{13}H_{17}$, $CHBr^{81}Br$,
 $C_2H_5$-CO-O$CH_2$CH(-O-CO-$C_2H_5$)-

## m/z 174

$CH_2$=CHCH$_2$N(-CO-$CH_3$)phenyl-, HBr$^{81}$BrC-

## m/z 175

$\underline{C_7H_5OCl_2}$, 174.9717 $Cl_2$-phenyl-CH(OH)-,
 $Cl_2$-(HO-)phenyl-$CH_2$-, $Cl_2$-(HO-)($CH_3$-)phenyl-,
 Cl-phenyl-O$CH_2$-

$\underline{C_{13}H_{19}}$, 175.1486 $R_n$-phenyl, perhydropyrene
also $C_4Cl_2F_3$, phenyl-Si(Cl)$_2$-,
 $H_2$C=C(Cl)$CH_2$O-CO-$C_2H_4$-CO-, $C_{12}H_{15}O$

| m/z, comp | Substructure, neighbor | Prop Abnd Spcf |
|---|---|---|

m/z 176
___

$\underline{C_{12}H_{16}O}$, 176.1201 cyclohexyl-phenyl-O-

m/z 177
___

$\underline{C_{12}H_{17}O}$, 177.1279 $C_4H_9$-(HO-)phenyl-CH($CH_3$)-,
  $C_4H_9$-($CH_3$O-)phenyl-$CH_2$-
also $C_2H_2{}^{81}BrCl_2$, $C_2H_5$O-CO-phenyl-CO-, $C_6H_{10}OBr$, $CF_2I$

m/z 178
___

$\underline{C_{14}H_{10}}$, 178.0782 dihydroethanoanthracene,
  -(phenyl)$_2$-$C_2$-
also $C_3Cl_3{}^{37}Cl$, HO-phenyl-N($C_4H_9$)$CH_2$-,
  phenyl-CH($CH_3$)N($C_2H_4$OH)$CH_2$-

m/z 179
___

$\underline{C_3Cl_3F_2}$, 178.9034

$\underline{C_{14}H_{11}}$, 179.0860 (phenyl)$_2$-$C_2$H-
also $C_2BrF_4$, $CH_3$O-CO-$C_3H_5$Br-, $C_2H{}^{81}BrClF_2$, $C_3HCl_3{}^{37}Cl$,
  $C_{11}H_{15}O_2$, $Cl_3$-phenyl-

m/z 180
___

$\underline{C_9H_{10}NO_3}$, 180.0660 $O_2$N-(HO-)phenyl-C($CH_3$)$_2$-

$\underline{C_{14}H_{12}}$, 180.0938 $CH_3$-phenyl-$CH_2$-phenyl-,
  $C_2H_5$-(phenyl-)$_2$-, $CH_3$-(phenyl-)$_2$CH-, -(phenylCH-)$_2$-

| m/z, comp | Substructure, neighbor | Prop Abnd Spcf |
|-----------|------------------------|-----------------|

---

## m/z 181

$C_4F_7$, 180.9888; phenyl-phenyl-CH($CH_3$)-,
Cl($CH_3H_7$-)phenyl-CH($CH_3$)-, phenyl-phenyl-NO-

---

## m/z 182

(phenyl)$_2$NCH$_2$-, ($O_2$N)$_2$-phenyl-NH-,
phenyl-CH$_2$-phenyl-NH-

---

## m/z 183

$\underline{C_7H_4BrO}$, 182.9446 Br-phenyl-CO-

$\underline{C_8H_8Br}$, 182.9810 Br-phenyl-CH($CH_3$)-

$\underline{C_{13}H_{11}O}$, 183.0809 phenyl-O-phenyl-CH$_2$-, $CH_3$O-(phenyl)$_2$-,
phenyl-CH$_2$-phenyl-O-, (phenyl)$_2$-O-CH$_2$-
also (phenyl)$_2$SiH-, $CF_3$SSCF$_2$-

---

## m/z 184

phenyl-CH$_2$-phenyl-O-, $C_{12}H_{26}N$, ($O_2$N)$_2$phenyl-O-,
$H_2$N-(Br-)phenyl-, -ON-phenyl-CO-CH2-

---

## m/z 185

$C_4H_9$O-CO-$C_4H_8$-CO-, $C_2Cl_3{}^{37}ClF$, -Cl-phenyl-OCH$_2$-CO-O-,
$C_3ClF_6$, $C_2HBr^{81}Br$, Br-(HO-)($CH_3$-)phenyl-

---

## m/z 186

phenyl-O-phenyl-O-, $C_4H_9$O-CO-$C_4H_8$-CO-, $C_2H_2Br^{81}Br$,
-Cl-(phenyl)$_2$-

MASS SPECTRAL CORRELATIONS

| m/z, comp | Substructure, neighbor | Prop Abnd Spcf |
|---|---|---|

<u>m/z 187</u>

$C_{14}H_{19}$, $Cl_2$-phenyl-C$(CH_3)_2$-, $C_2H_3Br^{81}Br$

<u>m/z 189</u>

$C_{14}H_{12}$, 189.1642; $C_8H_7OCl_2$, $C_{13}H_{17}O$,
  $CH_2$=CHCH$_2$O-CO-phenyl-CO-

<u>m/z 190</u>

Cl-($O_2$N-)phenyl-, NC-fluorene-

<u>m/z 191</u>

$\underline{C}_{15}\underline{H}_{11}$, 191.0860 anthracene-CH$_2$-, phenanthrene-CH$_2$-

$\underline{C}_{14}\underline{H}_{23}$, 191.1799 tetradecahydroanthracene-

$\underline{C}_{13}\underline{H}_{19}O$, 191.1435
also $C_4Cl_3F_2$, $C_3H_6ClO$-SiCl$_2$-, $C_3BrF_4$, $CBr^{81}BrF$,
  $C_4HCl_3^{37}Cl$, $C_3H_4^{81}BrCl_2$, $C_4ClF_4S$, $^{81}BrCl$-phenyl-,
  $C_2H_2F_2I$, $C_3H_7O$-CO-phenyl-CO-

<u>m/z 192</u>

$CH_3O$-phenyl-N$(C_4H_9)CH_2$-, -Cl$_2$-phenyl-CCl(-)-

<u>m/z 193</u>

$C_{15}H_{13}$, $Cl_3$-phenyl-CH$_2$-, $(C_2H_5O)_3SiOCH_2$-, $C_{12}H_{17}O_2$,
  $C_2H_5O$-CO-C$_3H_5$Br-, trimellitic anhydride esters

| m/z, comp | Substructure, neighbor | Prop | Abnd | Spcf |
|-----------|------------------------|------|------|------|

__m/z 194__

$C_4Cl_2F_4$, $C_2H_5O$-CO-phenyl-CO-O-, -($Cl_3$-)(HO-)phenyl-

__m/z 195__

$C_{15}H_{15}$, $Cl_3$-phenyl-O-, $CH_3$-phenyl-CO-phenyl-,
($CH_3$)$_2$-phenyl-O-P(=O)(-O-phenyl-$CH_3$)-, $C_{12}H_{16}Cl$,
$CF_3$-phenyl-$CF_2$-, Br-phenyl-$C_3H_3$-

__m/z 196__

$Cl_3$-phenyl-O-, phenyl-$CH_2$N(phenyl)$CH_2$-,
phenyl-($CH_3$-)-2-pyridonyl-

__m/z 197__

Br-phenyl-C($CH_3$)$_2$-, $C_2{}^{81}$BrClF$_3$, HO-phenyl-CO-phenyl-,
HO-($O_2$N-)$_2$phenyl-$CH_2$-, $CH_3$-phenyl-O-$CH_2$-, $C_4ClF_6$,
$C_4H_9$-(Cl-)phenyl-O-$CH_2$-, Br-CO-$C_2H_3{}^{81}$Br-,
(phenyl)$_2$Si($CH_3$)-, $C_{15}H_{17}$

__m/z 198__

($C_6H_{13}$)$_2$NCH$_2$-, HO-($O_2$N-)$_2$phenylCH$_2$-,
($CH_3$)$_2$N-CO-NH-(Cl-)phenyl-

__m/z 199__

$C_3H_3Br^{81}Br$, Br-(HO-)phenyl-CO-, $Cl_2$-benzofuran-$CH_2$-,
$C_3HCl_3F_3$

MASS SPECTRAL CORRELATIONS

| m/z, comp | Substructure, neighbor | Prop | Abnd | Spcf |
|---|---|---|---|---|

**m/z 200**

$-Cl-(CH_3-)(phenyl)_2-$, $Y*-C_3H_4Br^{81}Br-$

**m/z 201**

$C_3H_5Br^{81}Br$, $C_2HOBr^{81}Br$, $Cl$-phenyl-$CH_2$-phenyl-, $C_3Cl_2F_5$,
$C_2Cl_4{}^{37}Cl$, $(C_4H_9)_2(CH_3)_2Si_2H-$, $C_{16}H_9$

**m/z 202**

$Hg$, $-(Cl-)$phenyl-O-phenyl-, $Cl_2$-ext-ar-$OCH_3$

**m/z 203**

$Cl_2(CH_3)_3$-phenyl-O-, $Cl$-phenyl-O-phenyl-,
$Cl$-(HO-)phenyl-phenyl-, $Cl$-phenyl-phenyl-O-,
$C_{15}H_{23}$ (B/C/D rings of cholestane)

**m/z 204**

phenanthrene-cyc hc, $C_4H_5{}^{81}BrCl_2$, TMS dvts of
pyranosides

**m/z 205**

phenyl -O-/-OH, sesquiterpenones, $C_7H_5{}^{81}BrCl$,
$C_4H_9O-CO$-phenyl-CO-, $C_6F_7$, $C_2H_5O-C_6H_9Br-$,
$Br$-phenyl-$CF_2-$, $Br$-naphthyl-

**m/z 206**

$C_4H_9-(C_3H_5O-)$phenyl-O-, $Cl_3$-phenyl-$CH(-)CH_2-$

| m/z, comp | Substructure, neighbor | Prop Abnd Spcf |
|---|---|---|

__m/z 207__

phenyl C=O/-O-/-OH, $Cl_3$-phenyl-CH($CH_3$)-,
  methysiloxanes, $CBr^{81}BrCl$, $C_{16}H_{15}$

__m/z 208__

Pb, 207.9766; OCN-phenyl-$CH_2$-phenyl-,
  phenyl-phenyl-$OC_3H_3$-

__m/z 209__

R-(Cl-)phenyl, $C_{12}H_{14}ClO$, $C_4Cl_3{}^{37}ClF$, $C_{15}H_{13}O$,
  $CH_2$-(phenyl-CO-phenyl)-$CH_2$-, $Cl_3$-(HO-)-phenyl-,
  $C_3{}^{81}BrClF_3$, $C_3H_7O$-CO-phenyl-CO-O-

__m/z 210__

$O_2N$-phenyl-CH=CCl-CO-

__m/z 211__

HO-phenyl-phenyl-C($CH_3$)$_2$-, HO-($O_2N$-)$_2$phenyl-CH($CH_3$)-,
  $C_3HBrF_5$, $(C_4H_9O)_2P(=O)O$-, $C_2H^{81}BrCl_3$, $C_4HCl_3F_3$,
  phenyl-phenyl-Si($CH_3$)$_2O$-, $C_5^-H_{11}$-naphthyl-$CH_2$-

__m/z 212__

$C_3H_5$-(Br-)phenyl-O-, phenyl-O-phenyl-$C_3H_6$-

__m/z 213__

$C_4Cl_2F_5$, HO-(Br-)phenyl-C($CH_3$)$_2$-, $C_3Cl_4{}^{37}Cl$,
  $C_2{}^{81}BrCl_2F_2$, $C_4H_5Br^{81}Br$, $(CF_3)_2$-phenyl-,

| m/z, comp | Substructure, neighbor | Prop Abnd Spcf |
|-----------|------------------------|----------------|

$Cl(phenyl)_2$-CH=CH-, $\Delta$-5-(HO-) steroid A/B/C rings,
$C_6H_{13}$O-CO-$C_4H_8$-CO-

m/z 215
___

$C_{16}H_{23}$:Y*-steroid A/B/C rings, $C_4H_7Br^{81}Br$,
   $Cl_2$-$(C_2H_5-)_2$phenyl-$CH_2$-, $Cl_3{}^{81}Cl$-phenyl-,
   $CH_3$-benzanthracenyl-

m/z 217
___

$C_{16}H_{25}$:steroid A/B/C rings, R-phenyl,
   $Cl_2$-$(CH_3O-)$phenyl-$C(CH_3)_2$-, $Cl$-$(CH_3O-)$phenyl-phenyl-,
   $CH_3Hg$-, $(phenyl-O)_2P$-, $C_3Cl_3F_4$, TMS dvts of
   furanosides

m/z 218
___

I-(HO-)phenyl-, $(C_4H_9)_2$NCH(phenyl)-

m/z 219
___

$C_2Br^{81}BrCl$, $C_4F_9$, -$Cl_2$-phenyl-$OCH_2$-CO-O-,
   Br-phenyl-$SO_2$-, $^{81}BrCl$-phenyl-R-

m/z 220
___

Cl-phenyl-O-phenyl-O-, $(CH_3)_2$N-CO-$(C_4H_9-)$phenyl-,
   $C_4H_9$-(cyclohexyl)$_2$-

m/z 221
___

$C_6H_{17}O_3Si_3$, $(phenyl)_2$-$C_5H_7$-, $Cl_2$-$(phenyl)_2$-,
   $(H_3CO$-CO-$)_2$phenyl-CO-, $C_4ClF_6$, $Cl_2$-$(CH_3O-)_2$phenyl-O-,

| m/z, comp | Substructure, neighbor | Prop Abnd Spfc |
|---|---|---|

$Cl_3-(C_2H_5-)$phenyl-$CH_2-$, phenyl-CO-CH=C(phenyl)$CH_2-$,
$Cl_3-2,3$-dihydrobenzofuryl-, $Cl_2$-phenyl-$OCH_2$-CO-O-,
$C_2H_2Br^{81}BrCl$

m/z 222

$C_4BrF_5$, $C_5Cl_3F_3$

m/z 223

$CH_3Pb$, $C_3H_7$-(phenyl)$_2$-CH($CH_3$)-,
$Cl-(C_3H_7-)_2$phenyl-CH($CH_3$)-, $C_4H_9-(O_2N-)_2$phenyl-,
$C_4H_9$O-CO-phenyl-CO-O-, $C_4ClF_4S_2$,
$-(C_3H_7-)$(phenyl)$_2$-CH($CH_3$)-, $Cl_3$-phenyl-OCH($CH_3$)-,
$C_{10}H_{21}$-cyclohexyl-

m/z 224

$-C_{15}H_{30}$CH($C_5H_{11}$)-, $C_4H_9$O-(phenyl)$_2$-, $C_6F_8$

m/z 225

$Cl-(C_4H_9-)$(HO)phenyl-C($CH_3$)$_2$-, $C_4Cl_4{}^{37}Cl$, $C_3{}^{81}BrCl_2F_2$,
$C_5H_{11}$-naphthyl-CH($CH_3$)-, HO-$(O_2N-)_2$phenyl-C($CH_3$)$_2$-,
Br-$(C_3H_7-)$phenyl-CH($CH_3$)-, phenyl-O-CO-phenyl-CO-,
phenyl-O-phenyl-$C_4H_8-$, $C_5Cl_2F_5$, $C_{15}H_{31}$CH($C_5H_{11}$)-

m/z 226

$(C_7H_{15})_2NCH_2-$, $C_{18}H_{10}$, (phenyl)$_2$-O-$C_3H_4$(OH)-

| m/z, comp | Substructure, neighbor | Prop Abnd Spcf |
|---|---|---|

___

m/z 227

___

$C_{18}H_{11}$-, $C_4H_9$-(HO-)(Br-)phenyl-

___

m/z 228

___

$C_{18}H_{12}$, $-C_5H_8Br^{81}Br$-

___

m/z 229

___

$Cl_3{}^{37}Cl$-($CH_3$-)phenyl-, $C_5H_9Br^{81}Br$-, $C_3H^{81}BrClF_4$,
BrHC=$C^{81}BrC(OH)(CH_3)$-, $C_{11}H_{13}Si_3$,
$Cl_3$-phenyl-O-Y-$^{37}Cl$, $C_4HCl_3{}^{37}ClF_2$, $C_4Cl_3F_4$,
(phenyl)$_3$-, $F_3CCClFC(CF_3)(OCH_3)$-

___

m/z 230

___

$C_6H_{13}$-CO-N(phenyl)-$C_3H_4$-

___

m/z 231

___

$(C_4H_9)_2$-phenyl-C($CH_3)_2$-, $ClC_2H_4O$-(Cl-)phenyl-C($CH_3)_2$-,
phenyl-CH($C_{10}H_{21}$)-, perhydrobenzanthracene-,
$ClC_3H_4O$-CO-$C_2H_4$-CO-$OC_3H_4$-, $C_2H_5Hg$-, $C_5F_9$,
$ICH_2CH$(phenyl)-, $Cl_3{}^{37}Cl$-(HO-)phenyl-

___

m/z 233

___

disbustd(HO-)steroid A/B/C rings, $Cl_3$-benzofuran-$CH_2$-,
$(ClC_2H_4O)_2POC_2H_4$-, Br(phenyl)$_2$-, $C_7ClF_6$,
(phenyl-O)$_2$P(=O)-

| m/z, comp | Substructure, neighbor | Prop | Abnd | Spcf |
|-----------|------------------------|------|------|------|

### m/z 234

$-Br^{81}Br$-phenyl-

### m/z 235

$(Cl$-phenyl$)_2CH-$, $C_3Cl_3{}^{37}ClF_3$, $Br_2$-phenyl-,
$ClC_3H_6O-Si(Cl_2)OC_2H_4-$, (cyclohexyl-$C_2H_4-)_2CH-$,
$Cl_3(C_2H_5)_2$-phenyl-

### m/z 236

$HO-(O_2N-)$phenyl-$C_7H_{14}-$, $-(Cl_2-)$-phenyl-O-phenyl-,
$CH_2=CH-S-(Cl_3-)$phenyl-$Cl_3(C_3H_5)$-phenyl-O-,
$-C_{12}H_{23}-(C_5H_9)-$, $-(Br_2-)(CH_3-)(HO-)$phenyl-, $C_7F_8$

### m/z 237

$C_2H_5Pb$, $C_8H_{17}$-thiophenyl-$C(CH_3)_2-$,
$Cl-(C_3H_7-)_2(CH_3-)$phenyl-$CH(CH_3)-$,
$Cl_2$-phenyl-O-phenyl-

### m/z 238

$(CH_3)_2Pb$

### m/z 239

$^{81}BrCl_3$-phenyl-$CH_2-$, $C_5H_{11}$-naphthyl-$C_3H_6-$,
phenyl-$SiH_2$-phenyl-$Si(CH_3)_2-$, $C_{19}H_{11}$, $C_{17}H_{35}$,
$C_4H_5{}^{81}BrCl_3$, phenyl-O-phenyl-$C_5H_{10}-$, $C_{15}H_{31}-CO-$

| m/z, comp | Substructure, neighbor | Prop Abnd Spcf |
|---|---|---|

---

### m/z 240

$C_5Cl_3{}^{37}ClF_2$, $C_{19}H_{12}$, $Cl_3$-phenyl-$OCH_2CH_2O$-

---

### m/z 241

$Br^{81}Br$-cyclohexyl-, $C_4BrF_6$, $C_5Cl_3F_4$,
$ClC_2H_4OC_2H_4O$-phenyl-$C(CH_3)_2$-, benzophenanthrene-$CH_2$-,
$^{81}BrCl_2$-(HO-)phenyl-, $C_4H_9O$-CO-$C_8H_{16}$-CO-,
$C_5HCl_3{}^{37}ClF_2$, CHBr=CBrC(OH)($C_2H_5$)-

---

### m/z 243

$Cl_2$-($C_2H_5$-)$_3$-phenyl-$CH_2$-, $Cl_2$CH-phenyl-$CCl^{37}Cl$-,
(phenyl)$_3$C-, $Cl_3$-phenyl-$SO_2$-, $C_6F_9$

---

### m/z 244

(phenyl)$_2$N-phenyl-

---

### m/z 245

tetrahydronaphthacene-$CH_2$-, perhydronaphthacene-,
($CF_3CH_2O$-)$_2$P(=O)-, $C_3H_5O$-($C_4H_9$)$_2$-phenyl-

---

### m/z 247

$Cl_3{}^{37}Cl$-(HO-)phenyl-O-, $C_5ClF_8$, $C_4Cl_3{}^{37}ClF_3$

---

### m/z 248

($C_2H_5$)$_4$-cyc-$Si_3O_3$-, -(phenyl)$_2$-$C_2Cl_2$-

| m/z, comp | Substructure, neighbor | Prop Abnd Spcf |
|-----------|------------------------|----------------|

### m/z 249

$Cl_3-(C_2H_5-)_2$-phenyl-$CH_2-$, $Br^{81}Br$-phenyl-$CH_2-$,
$(ClC_3H_6O-)_2SiCl-$, $C_6Cl_4{}^{37}Cl$, $C_3HCl_5{}^{37}Cl$,
$(ClC_2H_4O-)_2P(=O)OC_2H_4-$, $C_7Cl_2F_5$, (OCN-phenyl-)$_2$CH-

### m/z 250

$Br^{81}Br-(H_2N-)$phenyl-

### m/z 251

$CBr_2{}^{81}Br$, trisubstd ketosteroids,
$Cl-(C_4H_9-)_2$phenyl-$CH(CH_3)-$,
(phenyl)$_2$-$C(CH_3)_2CH_2C(CH_3)_2-$, $C_3Cl_4{}^{37}ClF_2$,
$Br^{81}Br-(HO-)$phenyl-$C_6H_{11}CH_2CH(C_{10}H_{21})-$

### m/z 253

$(CH_3)_3Pb$, $Cl_2$-phenyl-$C_2H_3{}^{81}Br-$

### m/z 254

$I_2$(iodine)

### m/z 255

$C_2HBr^{81}BrCl_2$, $CH_3$-benzanthracene-$CH_2-$

### m/z 256

$C_6Cl_2F_6$, $S_8$(255.7766, sulfur)

| m/z, comp | Substructure, neighbor | Prop Abnd Spcf |
|---|---|---|

**m/z 257**

$C_{19}H_{29}$, $Cl_3{}^{37}Cl-(C_2H_5-)$phenyl-$CH_2-$

**m/z 259**

$Cl_5{}^{37}Cl$-cyclohexyl, $C_5Cl_3{}^{37}ClF_3$, $C_4H_3{}^{81}BrCl_2F_3$,
$(CH_3-)_2$phenyl-$CH(C_{11}H_{21})-$,
$ClC_4H_8O-(Cl-)$phenyl-$C(CH_3)_2-$

**m/z 260**

$C_4Cl_5{}^{37}Cl$

**m/z 261**

$C_4Cl_3F_4S$, $C_5HCl_3F_5$

**m/z 262**

$-(Br{}^{81}Br-)$phenyl-$CO-$, $C_6F_{10}$

**m/z 263**

$Cl_4{}^{37}Cl$-phenyl-$CH_2-$, $Br{}^{81}Br$-phenyl-$CH(CH_3)-$, $C_2Br_2{}^{81}Br$

**m/z 264**

$C_2HBr_2{}^{81}Br$, $-(CH_3-)$cyclohexyl-$CH(C_{11}H_{23})-$

**m/z 265**

$C_2H_2Br_2{}^{81}Br$, $(C_3H_7-)_2$(phenyl)$_2-CH(CH_3)-$, $C_{17}H_{33}-CO-$,
$C_7H_{13}CH(C_{11}H_{23})-$

| m/z, comp | Substructure, neighbor | Prop Abnd Spcf |
|---|---|---|

---

m/z 266

$Cl_4{}^{37}Cl$-phenyl-O-, $(CF_3)_2$-triazine-$CF_2$-, $CI_2$

---

m/z 267

$C_4H_9$-phenyl-O-phenyl-$C(CH_3)_2$-,
$ClC_3H_6O$-($C_4H_9$-)phenyl-$C(CH_3)_2$-, $C_2H_5Pb(CH_3)_2$-, $CHI_2$,
$C_{17}H_{35}CO$-, $C_4Cl_3F_6$, (naphthyl)$_2$-CH-

---

m/z 268

$(C_8H_{17})_2$N-CO-, $C_8F_9$

---

m/z 269

$C_6H_4BrF_6$, $C_{19}H_{27}O$(Y*$_2$-hydroxyketosteroid),
$Br^{81}BrCl$-phenyl-, $C_{15}H_{21}Si_2$, $C_5F_{11}$

---

m/z 271

$C_{19}H_{27}O$(Y$_2$*-diketosteroid, $CF_3Hg$, $CF_3$-(phenyl)$_2$-$CF_2$-,
$Cl_3$-phenyl-O-phenyl-

---

m/z 272

$C_5Cl_5{}^{37}Cl$

---

m/z 273

$C_{19}H_{29}O$(Y*-hydroxysteroid), Br-($CF_3$-)phenyl-$CF_2$-,
$C_3HBr^{81}BrF_4$

| m/z, comp | Substructure, neighbor | Prop Abnd Spcf |
|---|---|---|

m/z 274
_____

$C_7F_{10}$

m/z 275
_____

$C_5Cl_4{}^{37}ClF_2$, $C_4H_9$-phenanthrenyl-C($CH_3$)$_2$-,
$C_4{}^{81}BrCl_2F_4$, $(CF_3CH_2O)_2P(=O)OCH_2$-,
$C_4H_9O$-CO-$CH_2O$-($Cl_2$-phenyl)-

m/z 276
_____

-($^{37}Cl$-)phenyl-$C_2Cl_4$-

m/z 277
_____

$CCl_2{}^{37}Cl$-phenyl-$CCl_2$-, $(C_2H_5)_5$-cyclotrisiloxane-,
$Br^{81}Br$-dihydrobenzofuryl-, $Br^{81}Br$-phenyl-C($CH_3$)$_2$-,
dibenzoanthracenyl-

m/z 278
_____

-$C_6H_{10}$-CH($C_{13}H_{27}$)-

m/z 279
_____

$C_3H_4Br_2{}^{81}Br$, naphthyl-CH($C_{10}H_{19}$)-, $C_5HCl_3{}^{37}ClF_4$

m/z 280
_____

-($C_9H_{18}$)CH($C_{10}H_{21}$)-

| m/z, comp | Substructure, neighbor | Prop | Abnd | Spcf |
|-----------|------------------------|------|------|------|

m/z 281

$(CH_3)_7Si_4O_4-$, $CH_3(C_2H_5)_2Pb$, $(C_3H_7-)_3(phenyl)_2-$,
$C_9H_5F_8O$, $C_6H_{13}$-phenyl-O-phenyl-$C_2H_4-$,
$C_{10}H_{21}CH(C_9H_{19})-$, $Cl_4{}^{37}Cl$-phenyl-S-, $C_6F_{11}$

m/z 283

$Br^{81}Br$-phenyl-CHCl-

m/z 284

$C_6Cl_5{}^{37}Cl$

m/z 285

tetrahydronaphthyl-$CH(C_{10}H_{21})-$, $C_8H_2F_9O$

m/z 286

phenyl-Bi-, $C_8F_{10}$

m/z 287

$C_{14}H_{29}CH(phenyl)-$

m/z 291

$C_5H^{81}BrClF_6$, decahydronaphthyl-$CH(C_{10}H_{21})-$

m/z 292

$Br^{81}BrC_3H_5$-phenyl-O-

| m/z, comp | Substructure, neighbor | Prop Abnd Spcf |
|---|---|---|

**m/z 293**

$Br^{81}Br-(HO-)phenyl-C(CH_3)_2-$,
  $C_4H_6Br_2{}^{81}Br-(HO-)phenyl-C(CH_3)_2-$, $C_4H_6Br_2{}^{81}Br$, $C_7F_{11}$,
  $(naphthyl)_2C=CHCH_2-$, $(ClC_3H_6O)_2SiCl-OC_2H_4-$

**m/z 294**

$Br^{81}Br-(C_3H_5-)phenyl-O-$, $-(C_{10}H_{20})-CH(C_{10}H_{21})-$

**m/z 295**

$Cl_3-phenyl-OP(=S)^{37}Cl-$, $C_6H_{13}-phenyl-O-phenyl-C_3H_6-$,
  $(C_2H_5)_3Pb-$, $(C_{10}H_{21})_2CH-$

**m/z 297**

$C_5Cl_3{}^{37}ClF_5$

**m/z 299**

$C_4H_9-pyrene-C(CH_3)_2-$

**m/z 301**

$C_2HBr^{81}Br_2Cl$

**m/z 305**

$C_8F_{11}$

**m/z 307**

$(C_5H_{11}-phenyl-)_2CH-$, $C_5Cl_6{}^{37}Cl$

| m/z, comp | Substructure, neighbor | Prop | Abnd | Spcf |
|---|---|---|---|---|

m/z 309

$C_{17}H_{35}CH(C_4H_9)-$

m/z 311

$Cl_3C-(Cl-)phenyl-CCl^{37}Cl-$

m/z 312

$C_7F_{12}$

m/z 313

$Br_2^{81}Br-phenyl-$, $C_5Cl_4^{37}ClF_4$

m/z 315

$C_3H_3Br^{81}Br_2Cl$

m/z 317

$C_9F_{11}$

m/z 319

$C_6Cl_6^{37}Cl$, $C_6F_{13}$

m/z 324

$(CF_3CH_2O)_2P(=O)OC_2HF_2(-)-$, $C_8F_{12}$

# MASS SPECTRAL CORRELATIONS

| <u>m/z, comp</u> | <u>Substructure, neighbor</u> | <u>Prop</u> <u>Abnd</u> <u>Spcf</u> |
|---|---|---|

<u>m/z 325</u>

$(HO-)(C_4H_9-)(phenyl-)_2C(CH_3)-$

<u>m/z 327</u>

$(C_4H_9O-CO-)_2C_3H_3-CO-OC_3H_6-$, $Br_2{}^{81}Br-phenyl-CH_2-$

<u>m/z 329</u>

$C_{17}H_{35}CH(phenyl)-$

<u>m/z 331</u>

$C_{21}H_{23}Si_2$, $C_7F_{13}$

<u>m/z 337</u>

$C_{21}H_{43}CH(C_2H_5)-$

<u>m/z 341</u>

$CCBr_2{}^{81}Br-phenyl-CH(CH_3)-$

<u>m/z 343</u>

$Br_2{}^{81}Br-(HO-)(CH_3-)phenyl-$, $C_8F_{13}$

<u>m/z 345</u>

$C_2HBr_2{}^{81}Br_2$, $I_2-(HO-)phenyl-$

<u>m/z 355</u>

silicones, $C_9F_{13}$

| m/z, comp | Substructure, neighbor | Prop Abnd Spcf |
|-----------|------------------------|----------------|

**m/z 359**

$C_3H_3Br_2{}^{81}Br_2$

**m/z 367**

$C_{10}F_{13}$

**m/z 368**

-cholestene-

**m/z 369**

$C_7F_{15}$

**m/z 370**

-cholestane-

**m/z 381**

$C_8F_{15}$

**above m/z 400**

m/z 405, $C_{10}F_{15}$; m/z 412, $C_9F_{16}$; m/z 417, $C_{11}F_{15}$;
  m/z 424, $C_{10}F_{16}$; m/z 429, silicones; m/z 431, $C_9F_{15}$;
  m/z 436, $C_{11}F_{16}$; m/z 443, $C_{10}F_{17}$; m/z 447,
  $C_4Br_3{}^{81}Br_2$; m/z 448, $C_{12}F_{16}$; m/z 455, $C_{11}F_{17}$;
  m/z 462, $C_{10}F_{18}$; m/z 467, $C_{12}F_{17}$; m/z 469, $C_9F_{19}$;
  m/z 474, $C_{11}F_{18}$; m/z 481, $C_{10}F_{19}$; m/z 486, $C_{12}F_{18}$;
  m/z 493, $C_{11}F_{19}$; m/z 505, $C_{12}F_{19}$; m/z 512, $C_{11}F_{20}$;
  m/z 517, $C_{13}F_{19}$; m/z 524, $C_{12}F_{20}$; m/z 531, $C_{11}F_{21}$;

m/z 536, $C_{13}F_{25}$; m/z 543, $C_{12}F_{12}$; m/z 555, $C_{13}F_{21}$;
m/z 562, $C_{12}F_{22}$; m/z 567, $C_{14}F_{21}$; m/z 574, $C_{13}F_{22}$;
m/z 581, $C_{12}F_{23}$; m/z 593; $C_{13}F_{23}$; m/z 605, $C_{14}F_{23}$;
m/z 617, $C_{15}F_{23}$; m/z 631, $C_{13}F_{25}$; m/z 643, $C_{14}F_{25}$.

RECEIVED October 28, 1981.

Elements typeset by Service Composition Co., Baltimore, MD.
Printed and bound by the Maple Press Company, York, PA.

Jacket design by Kathleen Schaner.
Production by Robin Giroux and Karen Gray.